Feng Shui with a Twist

"Diamonds are a girl's best friend," but not just because they're beautiful. Diamonds and other gemstones have been used throughout history for their powers of attraction, protection, and healing. In our modern world, watches and other devices use the vibrational frequencies of quartz crystals.

The ancient Chinese art of feng shui also uses crystals to remedy problems and enhance specific areas of our lives. *Gemstone Feng Shui* goes beyond the use of crystals in traditional feng shui to include other potent gemstones as well.

Author Sandra Kynes shows you how to use gemstones and feng shui to attract positive energy and counteract negative energy in your living space. Part One introduces you to gemstones and feng shui, teaches you how to use traditional feng shui tools, and instructs you on using gemstones to activate specific areas of your living space to change your life. Part Two includes a compendium of sixty-three different gemstones that will help you choose the right stones to realize your desires.

Read this book, and you'll be searching your jewel box for gemstones that you can put to work in your home or office.

About the Author

Sandra Kynes is an artist and explorer of Goddess-worshiping cultures. She has created and teaches the workshop "Understanding the Language of the Goddess," which is based on the work of Marija Gimbutas. Kynes's work with crystals and gemstones led her to devise the nontraditional application of gemstone feng shui. Her writings have been featured in Llewellyn's *Magical Almanacs* and Llewellyn's *Spell-a-Day Calendars* under the name Sedwin.

To Write to the Author

If you wish to contact the author or would like more information about this book, please write to the author in care of Llewellyn Worldwide and we will forward your request. Both the author and publisher appreciate hearing from you and learning of your enjoyment of this book and how it has helped you. Llewellyn Worldwide cannot guarantee that every letter written to the author can be answered, but all will be forwarded. Please write to:

Sandra Kynes
℅ Llewellyn Worldwide
P.O. Box 64383, Dept. 0-7387-0219-6
St. Paul, MN 55164-0383, U.S.A.

Please enclose a self-addressed stamped envelope for reply, or $1.00 to cover costs. If outside U.S.A., enclose international postal reply coupon.

Many of Llewellyn's authors have websites with additional information and resources. For more information, please visit our website at:
http://www.llewellyn.com

GEMSTONE FENG SHUI

CREATING HARMONY IN HOME & OFFICE

SANDRA KYNES

2002
Llewellyn Publications
St. Paul, Minnesota 55164-0383, U.S.A.

First Edition
First Printing, 2002

Book design by Donna Burch
Cover art © 2002 by Photodisc and Aztech New Media Corp.
Cover design by Kevin R. Brown
Editing by Joanna Willis
Interior illustrations by Llewellyn art department

Library of Congress Cataloging-in-Publication Data
Kynes, Sandra, 1950–
Gemstone feng shui: creating harmony in home & office / Sandra Kynes.—1st ed.
p. cm.
Includes bibliographical references and index.
ISBN 0-7387-0219-6
1. Feng shui. 2. Gems—Miscellanea. 3. Crystals—Miscellanea. I. Title.

BF1779.F4 .K95 2002
133.3'337—dc21

2002016024

Llewellyn Worldwide does not participate in, endorse, or have any authority or responsibility concerning private business transactions between our authors and the public.

All mail addressed to the author is forwarded but the publisher cannot, unless specifically instructed by the author, give out an address or phone number.

Any Internet references contained in this work are current at publication time, but the publisher cannot guarantee that a specific location will continue to be maintained. Please refer to the publisher's website for links to authors' websites and other sources.

Note: The practices and techniques described in this book should not be used as an alternative to professional medical treatment. This book does not attempt to give any medical diagnosis, treatment, prescription, or suggestion for medication in relation to any human disease, pain, injury, deformity, or physical condition. The author and publisher of this book are not responsible in any manner whatsoever for any injury that may occur through following the information contained herein.

Llewellyn Publications
A Division of Llewellyn Worldwide, Ltd.
P.O. Box 64383, Dept. 0-7387-0219-6
St. Paul, MN 55164-0383, U.S.A.
www.llewellyn.com

Printed in the United States of America

This book is dedicated to my family—
Jessie Gallagher, Janet Gahring, and Lyle Koehnlein—
for all their love and encouragement.

Contents

Introduction

Humans have been attracted to gemstones since prehistoric times. While beauty plays a part in the appeal of stones (who isn't dazzled by a radiant gem?), a stone's energy also has an impact on us.

Just as humans feel connected to other creatures, so, too, are we connected to Earth's plants and rocks. The ancient people of Europe felt a strong connection with the Earth's energy through stone so they built temples and other structures on *ley lines*—meridians of energy that envelop the Earth. Today we marvel at the ingenuity and purpose behind Stonehenge and the Avebury Circle (England), the alignments at Carnac (France), and the long barrow of Newgrange (Ireland).

Gemstones also have their place in astrology. Besides the use of birthstones, which has been popular in mainstream Western culture for years, gemstones also have correspondences to the days of the week, hours of the day, and even guardian angels. The origins of these correspondences date back centuries.

As is mentioned in many parts of this book, crystals have been used for healing in many cultures throughout history. In recent years we have looked to the past for wisdom and have rediscovered the curative energies abundant in nature in herbs, aromas, and crystals. As our ancestors knew and we have relearned, disease may not always originate in the physical body. Our subtle bodies (energy fields or auras) may encounter problems or become unbalanced, which can manifest into the physical body. Because of their nonphysical origins, these are difficult to remedy. Since crystals emit vibrations that affect the flow of energy, they are perfect for resolving problems in the subtle body. Even the modern medical establishment is beginning to accept and work with alternative therapies in combination with conventional treatment to confront illness on various levels.

Industrially, the vibrational frequencies of crystals have been utilized in watches and other devices for years. Gems and crystals are also used in telephones, televisions, and radios. Silicon chips run our computers and other appliances, and rubies are used in lasers. Gemstones and crystals have been an integral part of our lives without our being aware of their roles.

Traditional feng shui is a holistic approach to living. It strives to balance the energies of our home and work environments while trying to maintain a balance with the natural world. As a form of geomancy, feng shui draws on Chinese astrology as well as the *I Ching (Book of Changes)*, which was written approximately three thousand years ago. You can delve into feng shui to incredible depths for divination and guidance, or simply apply basic principles to create a healthy environment in which to cultivate a particular aspect of your life.

If you are already using feng shui or have done crystal therapy, you are aware of the power of gemstones. Gemstone feng shui simply applies gemstone and/or crystal therapy to one's environment through feng shui techniques. It is not meant to replace traditional feng shui, but adapts the use of several tools to invoke change. Gemstone feng shui can be applied on its own or utilized to supplement a traditional feng shui practice.

Gemstone feng shui focuses on the properties, energies, and colors of gems and crystals to help you invite change into your life. This book will show you how to apply some of the principles of feng shui combined with the wisdom of gemstones to help you find balance in your life and connection with the Earth.

PART ONE

Connecting with Earth's Energy

The Earth is our home, our mother. Even though our modern lifestyles have taken away much of our contact with the Earth, we cannot escape the influence that the Earth's energy has on us. Created by dynamic forces, crystals and gemstones carry the Earth's energy and can be used to draw on the vitality of her healing and balancing forces. While crystal therapy provides the means to tap into crystal and gemstone energy, the ancient art of feng shui provides the framework to apply their power to our homes and environments.

CHAPTER ONE

Gemstones

Minerals, Crystals, and Rocks

Nothing could be truer than the old adage: beauty is in the eye of the beholder. The main attraction of gemstones is their color and splendid features. What we see when we look at a gemstone is a light wave that has been altered by the internal structure of the stone. Our eyes register different colors according to the length of a light wave. The chemical and structural attributes of a gemstone alters the white light that passes through it. Some light waves are refracted while others are absorbed. The higher the *refractive index* (angle of deflected light), the more spectacular the sparkle. This is a result of how light enters, passes through, and then exits the stone (Figure 1.1).

When light enters a stone, it does not pass straight through, it is deflected. In a stone that has double refraction, the light that enters is split and each ray is refracted, or bent, at different angles. This causes the facets of a stone (such as zircon) to appear doubled. Double refraction is also responsible for *pleochroism*, an effect whereby a stone appears as a different color when viewed from different angles.

Not only do the chemical and mineral composition of stones create their color, but they also create special effects. Stones with these effects are referred to as *phenomenal stones.* The "cat's eye" is one of the most dramatic effects. Another is the "star" that can be

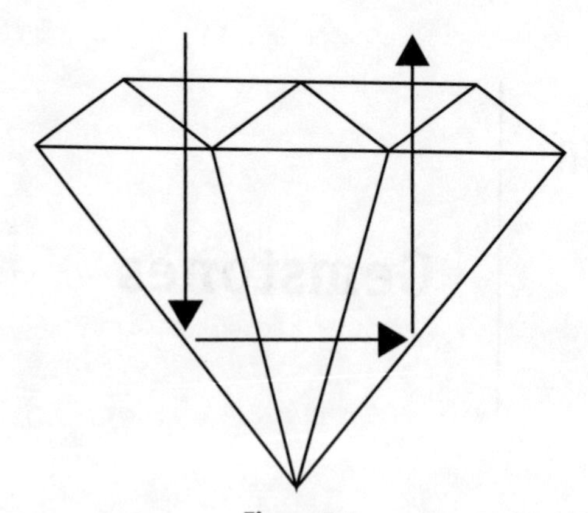

Figure 1.1
The refraction of light in a gemstone is what makes it sparkle.

found in sapphires, rubies, and a few others. Some gemstones produce a color change phenomena. Alexandrite, for example, may be green when viewed in daylight, but red under an electric incandescent bulb. Other gemstones exhibit a pleochroic effect. A stone that is dichroic will exhibit two different colors. A stone that is trichroic will show three colors.

Gemstones can be minerals or rocks. Minerals, a natural component of the earth, have a uniform, structurally separate chemical composition. Rocks are aggregates of disparate structures of one or multiple minerals. While there are several thousand types of minerals, only about a hundred are gemstones. Minerals can have the same chemical composition, however, varying conditions can produce very different structures. For example, diamond and graphite have almost the same carbon composition, but the conditions under which they are formed produce drastically different atomic structures and resulting appearances.

At its core, feng shui is a method of balancing the dynamic, ever-changing forces of the natural world. Gemstones are appropriate for feng shui work because they are formed through powerful, natural processes. The earth itself is dynamic and constantly changing. Minerals and rocks are continually being created, broken down, and re-created. As a result, gemstones embody the energy of the elemental cycles of creation and destruction that are utilized in feng shui.

The Earth has three basic layers: *core*, *mantle*, and *crust.* The core consists of two zones called the *inner* and *outer core*, and the mantle consists of an *upper* and *lower mantle* with a *transition zone* in between (Figure 1.2).

The Earth's crust is approximately 30 to 50 kilometers (19 to 31 miles) thick. Underneath the oceans it is only about 6 to 8 kilometers (4 to 5 miles) thick. The mantle, which is denser than the crust, is approximately 300 kilometers (186 miles) thick. The core, which has a radius of about 3,500 kilometers (2,175 miles), has a liquid layer—the outer core. The mantle slowly rotates around the core. In this intense pressure and heat, rock is melted

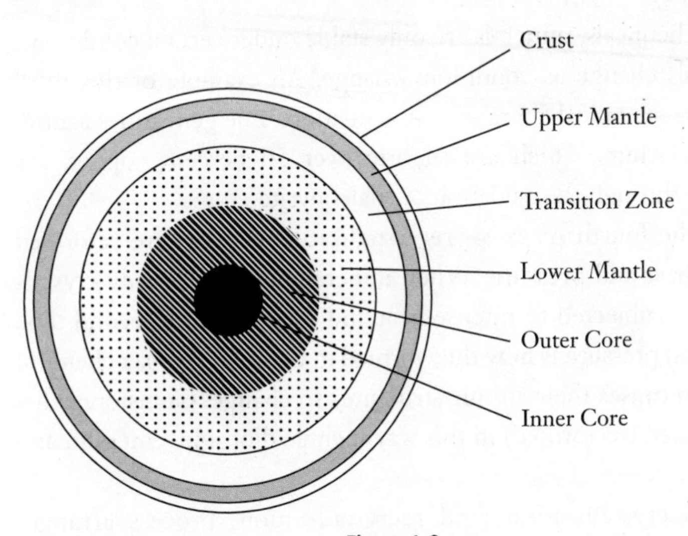

Figure 1.2
The Earth's layers.

and forms magma. As magma cools—either below the crust or on its way to the surface—it crystallizes.

Minerals are formed through four basic processes, and some are formed through a combination of these processes. They are (1) crystallization from molten rock, (2) precipitation from aqueous solutions, (3) chemical alteration, and (4) recrystallization.

In the first process—crystallization from molten rock—elements are combined under high temperatures. Minerals form as the elements cool. Examples of gemstones from this process include diamond, beryl, peridot, topaz, and zircon.

As the name of the second process suggests—precipitation from aqueous solutions—water is involved. Water dissolves and moves chemical elements from one place to another. During transport or at their new location, these elements interact with other chemical elements. Opal and gold are minerals that result from this process. Opal actually contains between 3 and 10 percent water.

As chemicals, minerals are only stable under certain conditions. Minerals change as conditions change. An example of this third process—chemical alteration—is oxidation. The gemstones azurite and malachite, which are slightly over 50 percent copper, are formed through the oxidation of chalcopyrite.

In the fourth process—recrystallization—atoms are reformed under heat and pressure. When minerals that are already crystallized are subjected to intense heat and pressure for a second time (heat and pressure is how they formed in the first place), a chemical reaction causes their atomic structures to change. Gemstones that are formed (re-formed) in this way include diopside, emerald, and kyanite.

The crystallization (and recrystallization) process arranges atoms into a particular structure. This structure is what determines a mineral's shape, color, and hardness. Even though minerals may have the same chemical composition, the crystal structure will create a different gemstone. This difference is initiated on the subatomic level. An atom that has become an ion (has an electric charge) has acquired or lost one or more electrons. This accumulating or repelling of electrons is responsible for different crystal structures.

Atomic structures create symmetrical exterior surfaces called *faces.* In order to form, crystals require conditions where they have room to grow. While there are over a hundred crystal shapes, they can be classified into seven categories according to their geometry. This geometry is based on the arrangement of faces around the center of the crystal. An imagined line through the center is called the *axis of symmetry.* How often the pattern of faces appears in one complete revolution of the stone defines its category. The categories are called *crystal systems.* The first system—isometric—is the most symmetrical. The other six decrease in symmetry (Table 1.1 and Figure 1.3).

These crystal systems were first developed in the late seventeenth century as the study of crystals, or crystallography, was undertaken. Until that time, crystals were believed to be some form of nonmelting ice. The word *crystal* comes from the Greek word *krystallos,* which in turn came from the word *kryos,* meaning "ice cold." This ancient theory is not completely far-fetched if you consider how a crystal—especially a large quartz crystal—feels in your hand. It seems to always remain cool to the touch.

Inclusions provide stones with additional characteristics and personality. Inclusions are macro- and microscopic imperfections within a crystal. They can be a solid, liquid, or gas. The cloudiness in some amber is caused by microscopic air bubbles. Inclusions can also be other minerals; rutile, for example, is frequently found in garnet and spinel.

Some gemstones are not created inside the Earth, but come from organic sources. Amber, as most people learned from the movie *Jurassic Park,* is fossilized sap from ancient pine trees.

Table 1.1—Crystal Systems

Category	Attributes	Examples
Isometric	Also called cubic; singly refractive	Diamond, fluorite, garnet, spinel
Hexagonal	Uniaxial; double refraction; can be dichroic	Aquamarine, beryl, emerald
Tetragonal	Uniaxial; double refraction; can be dichroic	Chalcopyrite, rutile, zircon
Trigonal	Uniaxial; double refraction; can be dichroic	Dioptase, quartz, ruby, sapphire, tourmaline
Orthorhombic	Biaxial; double refraction; can be trichroic	Alexandrite, chrysoberyl, peridot, tanzanite, topaz
Monoclinic	Biaxial; double refraction; can be trichroic	Jade, kunzite, malachite, moonstone
Triclinic	Biaxial; double refraction; can be trichroic	Labradorite, rhodonite, turquoise, sunstone

Tree-like coral comes from the Mediterranean Sea or Pacific Ocean. Pearls, which also come from the watery depths, begin as irritants inside oysters.

Gemstones Throughout History

People used stones, shells, and other objects for personal adornment as early as the Upper Paleolithic period (25,000–12,000 B.C.E.). Beads of carnelian and quartz date to 6000 B.C.E. in Mesopotamia.[1] Gemstone jewelry and talismans were found in

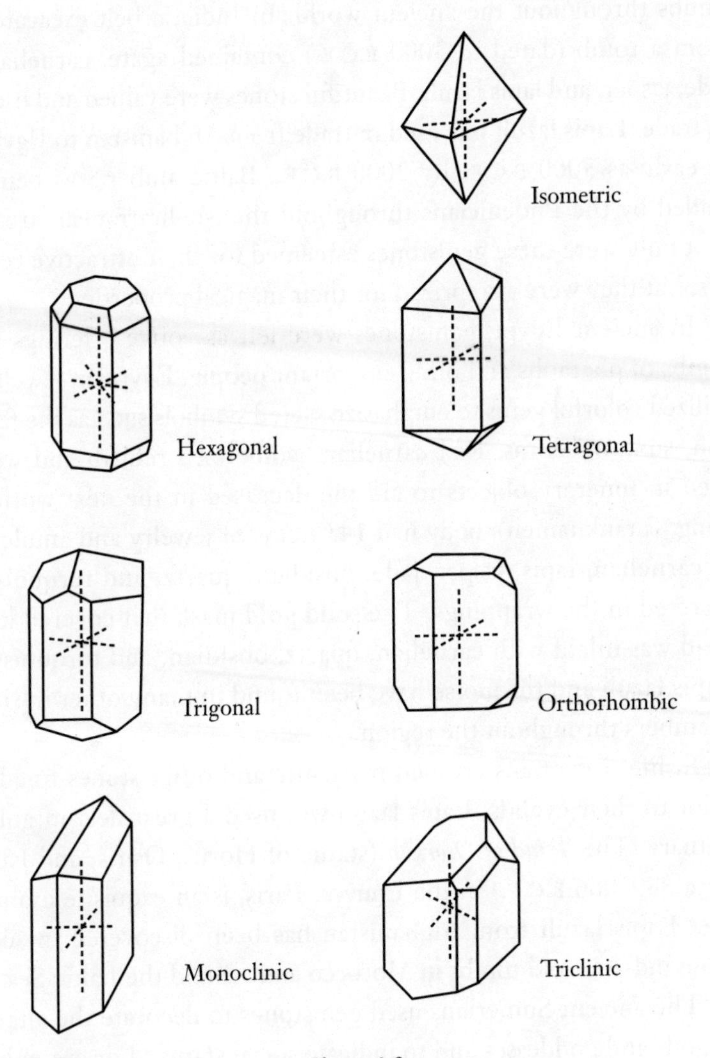

Figure 1.3
The crystal systems.

tombs throughout the ancient world. In India, a belt excavated from a tomb (dated to 3000 B.C.E.) contained agate, carnelian, jade, jasper, and lapis lazuli. Beautiful stones were valued and used in trade. Lapis lazuli was used in trade from Afghanistan to Egypt as early as 3000 B.C.E. By 2000 B.C.E., Baltic amber was being traded by the Phoenicians throughout the Mediterranean area.[2] Not only were these gemstones esteemed for their attractive colors, but they were also prized for their magical properties.

In ancient Egypt, gemstones were left as votive offerings in tombs of pharaohs and other important people. Egyptian jewelry utilized colorful gems to emphasize sacred symbols such as the falcon, sun, and lotus. Red carnelian symbolized rebirth and was used in funerary objects to aid the deceased in the next world. King Tutankhamen's body had 143 items of jewelry and amulets of carnelian, lapis, jasper, jade, obsidian, quartz, and turquoise secreted in the wrappings. The solid gold mask that covered his head was inlaid with carnelian, quartz, obsidian, and turquoise. Lapis lazuli and turquoise have been found in many other burial chambers throughout the region.

In life, Egyptians crushed malachite and other stones to add color to their eyelids. Lapis lazuli was used a great deal in gold statuary. The *Triad of Osorkon* (statue of Horus, Osiris, and Isis, circa 889–866 B.C.E.) in the Louvre, Paris, is an exquisite example.[3] Lapis lazuli from Afghanistan has been discovered in six-thousand-year-old tombs in Morocco and around the Baltic Sea.

The ancient Sumerians used gemstones to decorate the altars of gods and goddesses and to indicate social status.[4] For example, they used onyx and quartz for seals of state. These seals were carefully shaped and carved and occasionally were pierced, providing double duty as a necklace. Many other stones such as agate, carnelian, jasper, and obsidian were carved into figurines, cups, beads, and an array of objects for personal adornment.

The use of jade in China has been traced back four thousand years. The workmanship applied through carving increased the value of jade objects. Across the ocean in the New World, ancient Mexicans placed a higher value on jade and turquoise than on gold.

Gemstones are mentioned throughout the Bible. It mentions that the vestments worn by ancient high priests included a breastplate of twelve gemstones. These were believed to increase the priest's power during sacred ceremonies. Ezekiel 28:12–14 describes Hiram of Tyre's robe: "Thou hast been in Eden the garden of God; every precious stone was thy covering, the sardius, topaz, and the diamond, the beryl, the onyx, and the jasper, the sapphire, the emerald, and the carbuncle, and gold: the workmanship of thy tabrets and of thy pipes was prepared in thee in the day that thou wast created." (*Carbuncle* was a catch-all term for rubies, spinels, and garnets.) In Revelation 21:18–20: "And the foundations of the wall of the city were garnished with all manner of precious stones. The first foundation was jasper; the second, sapphire; the third, a chalcedony; the fourth, an emerald." Exodus 28:8–10 gives notice of early engraving: "And thou shalt take two onyx stones, and grave on them the names of the children of Israel."

The Christian practice of honoring the memory of saints by creating reliquaries with their remains called for the use of many jewels. The famous talisman of Charlemagne was decorated with two large sapphires and was said to hold a piece of the True Cross. Elaborate ornamentation in churches and cathedrals utilized gemstones. In the chapel of St. Joseph, Westminster Cathedral, England, Iberian agate and Canadian onyx provide a backdrop for a *fleur de peche* Italian marble column.

The oldest known text about minerals was written by Theophrastus.[5] Pliny included data about gemstones in his writings on natural history.[6] Gemstones were written about in relation to travels by Marco Polo (thirteenth century) and Jean Baptiste Tavernier (seventeenth century).

The ancient methods used for fashioning gemstones consisted of smoothing and polishing them to bring out their colors. These methods came to be perfected in India. It was not until the Middle Ages that cutting gemstones came into practice. From Italy it spread to other parts of Europe. Over the centuries many types of cuts have been developed; some are special to a particular type of stone—such as diamond or emerald—to bring out its specific qualities. Other types of cuts become the fashion rage and then fade.

In addition to being admired for their beauty, gemstones have been put to industrial use. Feldspar was the secret ingredient in the clay used to manufacture Chinese porcelain. Without this, Europeans could not produce porcelain of equal high quality for several centuries. Gemstones also serve us in our modern world. As was already mentioned, the piezoelectric properties of quartz have made them useful in our wrist watches. *Piezo-* comes from the Greek word *piezein*, which means "to squeeze." Basically, squeezing or putting pressure on quartz crystals causes them to release an electrical charge. Some gemstones have pyroelectric properties—an electric charge is released when the stone is heated and cooled.

Other industrial uses of crystals include silicon chips, transistors, LEDs (light emitting diodes), tape recorder heads, gas igniters on stoves, LCDs (liquid crystal displays), diamond drills used by dentists, and carbide crystals that give steel its hardness.

---

1. Mesopotamia, circa 7000–500 B.C.E., occupied what is present-day Iraq and parts of Syria and Turkey. Around 550 B.C.E. it became part of the Persian Empire.
2. Sofianides and Harlow, *Gems & Crystals*, 20.
3. Graham, *Goddesses in Art*, 55.
4. The civilization of Sumer, located in southern Iraq where the Tigris and Euphrates Rivers meet, flourished approximately seven thousand years ago.

5. Theophrastus (circa 372–287 B.C.E.) was a Greek philosopher, naturalist, and devoted pupil of Aristotle.
6. Gaius Plinius Secundus (circa 23–79 C.E.), known as Pliny the Elder, was a Roman scholar and writer who, among many other books, compiled a thirty-seven-volume encyclopedia on natural history.

CHAPTER TWO

Crystals

Crystal Therapy

Our ancestors lived in a world illuminated only by the sun, moon, and fire or candle light. Gemstone energy helps us find that balancing primal light that kept early people in touch with the natural rhythms of life-force energy. Crystals are gifts from Mother Earth that carry her energy. This natural energy nourishes us.

Modern crystal therapy is not a New Age invention or a toy. The ancient Greeks ascribed special healing powers to gemstones. Many centuries later the bishop of Regensburg, Albertus Magnus (circa 1193–1280), wrote about the healing properties of gemstones in five volumes called *De Mineralibus et Rebus Metalicis*. Albertus Magnus was a German philosopher, theologian, and scholar whose work was widely respected.

Gemstone amulets were worn for particular ailments in medieval Europe. Gemstones were also ground up and included in curative potions. If at first this sounds far-fetched to you, stop and consider that modern antacids contain limestone, and an integral ingredient of Pepto-Bismol is the mineral bismuth.

The key to the healing powers of gemstones is their vibrations. Vibrations surround us all the time. Mostly we are aware of them as sound and light. Vibrations are carried throughout our bodies by electrical impulses that jump the synapses between our cells. This is part of our electromagnetic field and it extends outside our bodies to what is known as the *aura.* Crystal vibrations interact with the electromagnetic fields of our bodies and can enhance our personal energy.

Modern scientific research into the human energy field began in mid-nineteenth-century London when Dr. Walter Kilner became interested in the latest medical technology of x-rays and electrotherapy. He developed a process that utilized the ultraviolet spectrum of light and enabled him to view the aura energy field. In 1939, Semyon Kirlian, a Soviet engineer, developed a method for photographing the aura. Today it is almost commonplace to find an alternative bookshop offering aura photography and interpretation.

The aura is the outward manifestation of the body's energy. Inside, the body has seven specific energy points called *chakras* (Table 2.1). The chakras have been described as spinning or rotating wheels of energy. *Chakra* comes from the Sanskrit word for "wheel." Humans are creatures of light and energy; the seven chakras vibrate at different frequencies that match the frequencies of light passed through a prism. White light is composed of seven primary colors: red, orange, yellow, green, blue, indigo, and violet. When light is broken down as through a rainbow or cut-glass prism, we can see these seven colors. Red has the lowest and longest frequency and vibration.

Some practitioners include an eighth chakra called the *transpersonal point.* This is located a few inches above the head and connects us with the Divine. This chakra is also referred to as "heavenly *chi.*"

The Laying-on of Stones

One form of crystal therapy works directly with the chakras to realign the body's energy field. This method, called "the laying-on of stones," is performed by having a person lie on his or her back, then placing an appropriately colored gemstone in the area of the problem chakra. If there is no specific problem, this method can be used with all of the chakras to align energy flow. A general balancing or

Table 2.1—Chakra Associations

Color	Chakra	Physical Area	Aspect	Gems for Therapy
Red	One	Perineum/ base of the spine	Survival, security, safety	Red garnet, smoky quartz
Orange	Two	Adrenals/ abdomen	Strength, sexuality	Carnelian, tiger's-eye
Yellow	Three	Solar plexus/ stomach	Personal power, ego, impulses	Citrine, topaz
Green	Four	Cardiac plexus/ heart	Compassion, love	Rose quartz, kunzite, jade
Blue	Five	Thyroid gland/ throat	Creativity, speech, writing	Aquamarine, azurite
Indigo	Six	Pituitary gland/ forehead	Higher intuition, psychic abilities	Lapis lazuli, sapphire
Purple/ Violet	Seven	Pineal gland/ top of head	Spirituality, enlightenment	Amethyst, white or clear quartz

clearing of negative energy can be done by using clear quartz crystals. When working with the chakras, be sure to visualize each one separately as calm and bright before moving on to the next. This can also be combined with a Reiki session or massage for a powerful movement of energy.

When multiple stones are used on the body, lay them in the same direction with their points toward the head or toward the feet. These directions are known as sky-to-earth and earth-to-sky. Using crystals in this way can enhance the flow of energy. For stones that are not pointed, hold each one separately before starting the session until you get a sense for which side is "up." Trust your intuition. If your healing calls for energy to be grounded, place the stones pointing toward your feet. If the healing is spiritual or requires a "lifting" of emotions, point the stones toward the crown of your head.

Keep It Close

Another method of gemstone/crystal therapy is to simply wear or carry a stone that personifies the attributes you want to attain or stimulates energy for healing. In addition to finding the corresponding color and type of gemstone, an important aspect in selecting a particular stone is your attraction to it. See Appendix A for more on buying crystals and gemstones.

Gemstone/Crystal Meditation

In addition to crystal therapy, utilizing gemstones in meditation is another way to access a stone's energy. In meditation we close off the conscious area of our brains, which allows us to access the subconscious. Gemstones can be used to help. One method of gemstone meditation is to sit with your hands cupped in front of you holding your stone or crystal. If you are using two stones, sit with your hands on your knees, palms up, with a stone in each. Another method of tapping into a gemstone's energy is to gaze at it. Soften your focus as you look at the stone and let your eyes rest on the nuances of color and form. Crystals show amazing characteristics when viewed through the light of a candle. Interior cracks and frosts are revealed by the light of the flame. Allow the crystal to draw you in to share its gifts and special energy.

The art of crystal gazing (and scrying with mirrors made of obsidian) has been mentioned throughout English literature. Chaucer's knight in "The Squire's Tale" of *The Canterbury Tales* arrives with a mirror that allows those who look in it to see "the coming shadow of adversity," as well as discern who is friend or foe. Shakespeare also uses this device in *Measure for Measure* and *Macbeth*. A mirror that reveals a completely different view of the world figures largely in Lewis Carroll's *Through the Looking Glass*.

In order to do a gemstone/crystal meditation (or any type of meditation), be sure to allow yourself adequate time. If you have a gazillion things to take care of, don't try to squeeze a meditation in between other items on your agenda. It's important to allow time not only for the meditation itself, but also for reflection afterward. Taking a few minutes to jot down your thoughts and feelings after meditation is a good way to track the effects different gemstones may have on you. Schedule time in a place that is quiet and private. A loud television in the next room or kids bounding around the house would be a distraction even for experienced meditators.

If you have not meditated before, begin by sitting comfortably and closing your eyes. Even if you plan to crystal gaze, start with your eyes closed. This provides you with the opportunity to shift from the everyday outer world to your own interior space. Focus on your breathing and let each breath start from your belly. Slowly fill your lungs, then pause before you slowly exhale. The last air should leave from your belly. Pause again, and then start the next inhalation. When you feel that your own energy is calm and grounded, allow your focus to move to the gemstone(s). If you are crystal gazing, slowly open your eyes but keep the focus soft. Allow yourself to be receptive to energy, thoughts, messages, and feelings.

Don't approach the meditation with expectations of great, earth-shattering revelations about your life. Most information will come to you softly. And don't be disappointed if nothing seems to happen. Just allow yourself to relax and be receptive. If you are to learn something at that time, it will come.

Taking time after a mediation is important even if you do not keep track of your experience in a journal. Having time for reflection allows information to settle. Things that may not be obvious

Table 2.2—Gemstones for Western Elemental/Seasonal Energy Balance

Season	Element	Direction	Color	Gemstones
Spring	Air	East	Green	Emerald, agate, beryl, jade
Summer	Fire	South	Red	Ruby, garnet, carnelian, beryl
Autumn	Water	West	Blue	Sapphire, lapis, opal, turquoise
Winter	Earth	North	White	Diamond, pearl, quartz, spinel

Note: The first gemstone in each season is the traditional stone.

during the meditation may come to the surface while you sit quietly afterward. It may also take a day or two for you to realize any changes.

Another way of utilizing gemstones during meditation is to surround yourself with them. Set them out in a circle on the floor, then sit in the middle. The stones do not have to touch each other and can be spaced as widely apart as seems appropriate. Use the charts and other information in this book to select gemstones for specific purposes.

Western Elemental/Seasonal Energy Balance

In addition to working with the chakras for centering, you might want to try a simple elemental balancing to help you get grounded. Use Table 2.2 to coordinate your selection of stones. Choose the gemstone specified or one that is predominantly an elemental color. This balancing utilizes the four basic elements common in Western traditions. Chinese feng shui employs five elements, but that will be covered later.

Gather one gemstone for each of the seasons/elements and set it on the floor in its corresponding direction as though you were going to create a circle in which to sit. You can use just these four stones or complete the circle with either white or black stones. (White is the presence of all colors and black is the absence of all colors.) From inside the circle, face the direction that corresponds to the current season. If it is midway through a season, face the midway point.

Begin as you would with other meditations by focusing on your breath and calming your energy. With your thoughts relaxed, move clockwise and think of the first direction and element. In the case of the example in Figure 2.1, you would think of East, springtime, and the element air. Allow yourself to feel the energy of this season and experience its element. Let the energy of the stone guide you. Repeat this process for each of the four directions. Spend as much time as you feel is appropriate on each one. With this meditation you are connecting with Mother Earth and the entire web of life. This meditation can help you stay grounded for several days.

If you practice yoga, you might want to do the elemental/seasonal grounding meditation during a session at home. Place the four gemstones around your mat and do a specific posture that evokes the season or element for you. For example, you might do Downward-facing Dog for summer/South because the posture creates heat and warms the body. You might also try crystal gazing while doing yoga to deepen the meditative state.

Experiment with different forms of meditation or sitting with gemstones to find what is comfortable for you. Meditating, doing yoga, or just sitting with your gemstones will help you tune into the energy of each and aid you in selecting appropriate stones for practicing feng shui.

Preparing Gemstones for Use

Whenever we get new clothes or food from the store, we usually wash it before wearing or eating it to ensure that it is clean and not carrying something unwanted. The same goes for gemstones. Before using a gemstone, it should also be cleansed to remove

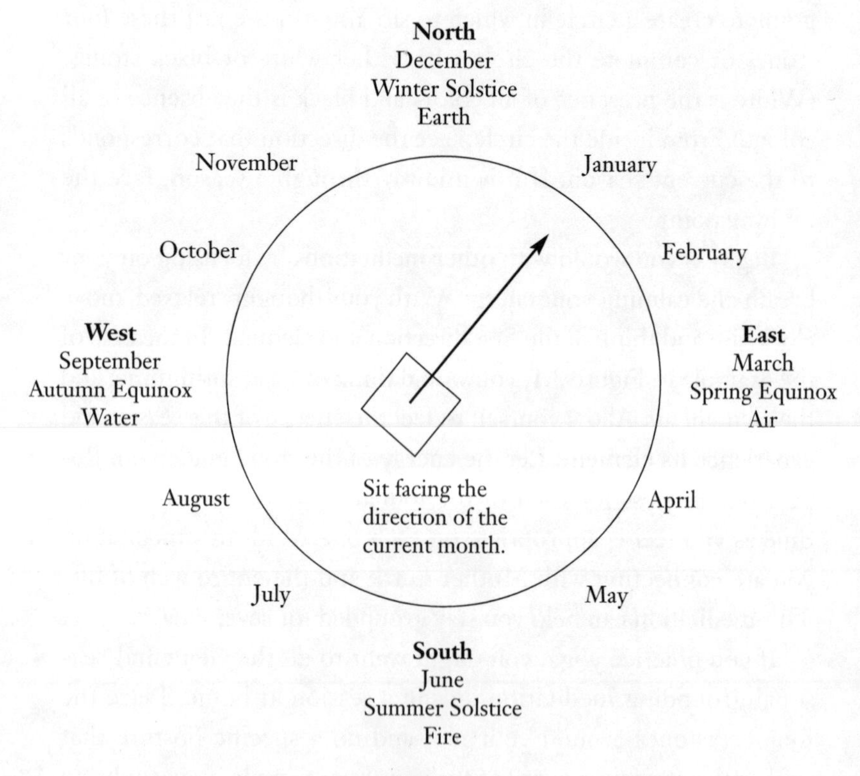

Figure 2.1
Western elemental/seasonal energy balance.

unwanted or negative energy that it may have picked up through previous handling. (Refer to Appendix A for information on acquiring gemstones.) Even if you are strongly attracted to a stone, you should cleanse it in order to allow the greatest amount of energy to flow between it and you. Cleaning it will allow a clear, pure flow. It will help you get the most power from the gemstone too. Over time you may feel that a stone has lost its potency, which indicates that it needs to be recleaned. Gemstones should be cleaned separately to keep their power focused on their own energy.

Salt Cleansing

Cleansing with salt is advocated especially for gemstones that are new in your life. Sea salt is traditionally recommended, however, this should be a personal decision. Because it comes from the ocean, sea salt is imbued with the power of the waves and the cleansing intention of water. If you find that you are a "North person" (see chapter 4 to determine your direction on the pa tzu compass), salt that is mined from the ground may be more appropriate for you. This salt comes from the body of Mother Earth, which will give the gemstone a strong power for grounding. Use pure salt; many types on the market contain aluminum or other chemicals.

Another decision with salt cleansing is whether to do it wet or dry. For wet cleansing, add a tablespoon of salt to a cup of water. You may start with warm water to dissolve the salt, but let it cool before bathing the gemstone. Water that is too hot may cause the stone to crack and fracture. Use a glass or ceramic container and avoid plastic or metal as these tend to leach some of their own properties into the water. Allow the gemstone to soak overnight, then dry it with a soft cloth. For dry cleansing you will need a container deep enough to fill with enough salt to bury the gemstone. Place the stone in the salt point (or top) down, facing Mother Earth, and leave it overnight.

Moonlight Cleansing

Another popular way to cleanse a gemstone is by moonlight. This takes a little longer, but if you want to activate the power of Luna it's worth the wait. Find a windowsill in your home that gets at least several hours of moonlight during the full moon. Place the gemstone on that windowsill for three nights beginning with the night before the full moon. Placing the stone on a porch or safe place outside is even better as the light is not impeded by window glass. To recharge the gemstone from time to time, use the light on the one night of the full moon.

If a gemstone has been used for clearing negativity, place it in a clear glass container of salt water and position it where it will receive the light of a waning moon for several nights. Luna will take the negativity with her as she goes away. A gemstone cleansed in salt water and placed outside (or on a windowsill) on the night of a new moon will be open to receive inner strength from Luna's darkness. The light of a waxing moon will help boost the power of a stone.

Herb/Flower Cleansing

A gentle method of cleansing a gemstone is to bury it in a bowl of dried herbs or flower petals. This can be combined with moonlight cleansing, but on its own it takes approximately a week. For extra grounding, bury the gemstone outside in the earth or inside in a cup of soil with its point or top downward.

A speedier way to cleanse a gemstone is through *smudging*. In a fireproof bowl or seashell, burn a little sage, cedar, or mugwort, and then pass the gemstone through the smoke. You don't need to generate thick smoke for this; a gentle wafting will do nicely. A few lavender flowers can be added for calming energy. Also be aware that mugwort has a similar odor to marijuana when burned; use this herb judiciously.

After cleansing a new gemstone, take time to sit with it. Cradle it in your hands and welcome it into your life. Be open to the energy of the stone and establish a connection to it with your own energy. Once you have done this, pass along your intentions for it. Think of how you want to share its energy in your home and call on its powers.

CHAPTER THREE

Basic Concepts of Feng Shui

Yin and Yang: Cosmic Balance

Feng shui literally translated is "wind and water," the two most prevalent elements that shape our world. Feng shui can be used to shape the energy around you. It is about changing your life by assessing and maintaining your connection to the natural world. The dynamic forces—*yin* and *yang*—need to be kept in balance in order to create change and achieve harmony and happiness (Table 3.1). Learning to balance your life means walking "between the magnetic fields of yin and yang."[1] In Zen Buddhism it is called going through "the gateless gate."

Table 3.1—Balance

Yin	Yang
Female	Male
Cold	Hot
Rest	Activity
Soft	Sharp
Land	Water
Winter	Summer
Night	Day
Moon	Sun
Goddess	God

Don't immediately assume that based on your gender everything yin or yang applies to you. Although these energies have nothing to do with gender, human culture has adopted the idea that to be male is to be active, dominating, aggressive, and analytical, and to be female is to be receptive, yielding, supporting, and intuitive. A person is not completely one or the other; women have male/yang energy and men have female/yin energy and it is important to balance them.

One way you can achieve yin/yang balance is to carry with you a stone of the energy opposite to what you want to tone down or the same as what you want to encourage. For example, if you feel you are being overanalytical, carry a stone that enhances yin energy (aquamarine, iolite, or moonstone). If you feel that you are too passive, use a stone that is yang (carnelian, garnet, or sardonyx). See Part II for more information.

Yin and yang can best be described as a harmonious dynamic of opposites. They are the binding forces that hold the universe together. They are present in all things and represent what author Johndennis Govert calls "the Ultimate Source." Because the forces of yin and yang are constantly changing, the universe is anything but static.

Water, which is basically yang, can be either yin or yang: a still pond gathers in energy and is yin, while a rushing stream sends out energy and is yang. The turning of night and day, as well as the seasons, reflects the balance of yin and yang. The spring and autumn equinoxes are times of balance when night and day, and winter and summer are equal.

In ancient China, yin feng shui was used to design grave sites. Great care was given to creating a burial place because it was believed that one's fortune was linked to how well their ancestors' graves had been prepared with the proper balance.[2] A yin dwelling is a place of burial. A yang dwelling is a place for the living.

Because yang energy is life energy, it should have dominance in the home. However, yin energy should not be excluded because

Figure 3.1
The yin and yang symbol.

it is necessary for balance. After all, life is but part of the cycle of life, death, and rebirth, and to experience and enjoy life to the fullest requires a balance of energy.

The yin and yang symbol (Figure 3.1) not only illustrates that a balance of opposites is necessary to create a whole, but also that even opposites contain a little of each other. Just as the sun is necessary for growth, the cool quiet of the night is necessary for rest. After the summer growing season, winter provides time for the earth to rest and renew. Yin and yang represent a continuous cycle of change. The symbol itself is very fluid and interactive.

Energy: The Breath of Life

Anyone who has taken a yoga, *t'ai chi*, or meditation class, or has been totally absorbed in a creative pursuit, has experienced the flow and use of energy. Increasing numbers of people are discovering the power of their energy and their connection with the natural world. Bringing gemstones into our lives helps to keep us close to nature and in tune with the greater web of life. Our modern technology has provided innumerable conveniences, but it has also insulated us from much of the natural world and its flow of vital energy. To be healthy, energy needs to be moving, but it must move gently and smoothly in order to activate the life force within us.

Meditation, yoga, t'ai chi, and other disciplines involve a shift away from ordinary consciousness in order to tap into the rich field of energy where a person is no longer a separate small entity but part of the larger universe. Starhawk has called this awareness *starlight vision.*[3] Maintaining an environment where healthy energy constantly flows around you helps to support spiritual and creative pursuits, as well as everyday well-being. It helps us connect with our starlight vision.

Healthy, active energy is called *chi*, *sheng chi*, or *qi*. This chi—positive energy—meanders slowly. It is most healthy where yin and yang are balanced and where there is elemental harmony. In order to stay positive, energy needs to drift and flow around the contours of buildings and the land itself. The lay of the land and vegetation affects how energy moves.

Manmade environments—towns, cities, highways—have a strong impact on energy not only in their immediate vicinity, but also in the surrounding areas. Urban areas can be more challenging because of their fast pace and the discordant effect of buildings that are erected without consideration of other structures. However, the natural landscape can also present problems. This is why it is important to look outside your home as well as within to assess energy patterns.

When Good Goes Wrong

Negative energy moves rapidly in straight lines. It occurs where yin and yang are out of balance or where elemental energy is generated through a destructive cycle. This negative chi is referred to as *sha* or *shar chi*, which is extreme to either end of the scale. When energy is trapped (e.g., in a corner or in a valley), it stagnates into sha that lacks the vitality of life and is too yin. Likewise, when chi is forced into a straight line, it gains momentum and becomes dangerous and threatening sha (too yang). Chi is the calm, balancing energy in between the two extremes of sha (Figure 3.2).

Because we usually do not have complete control over our external environments, the most important function of feng shui deals with disarming the negative energy that bombards us. While it is good to do as much as possible to enhance positive chi, it is more important to deflect and disperse the negativity that could be draining good things, including the life force, from your home. Check for negative energy first, determine a strategy to guard against it, then develop a plan for building and attracting positive energy.

Energy is affected by shape, color, smell, lighting, temperature, and all tangible objects. A densely populated area with fast-moving cars, glaring lights, and unpleasant smells jangles the nerves because of the uneven (fast, rough, spiky) energy that it creates. It is also possible to have too much of a good thing: despite an idyllic view, a room with excessively large or a great many windows

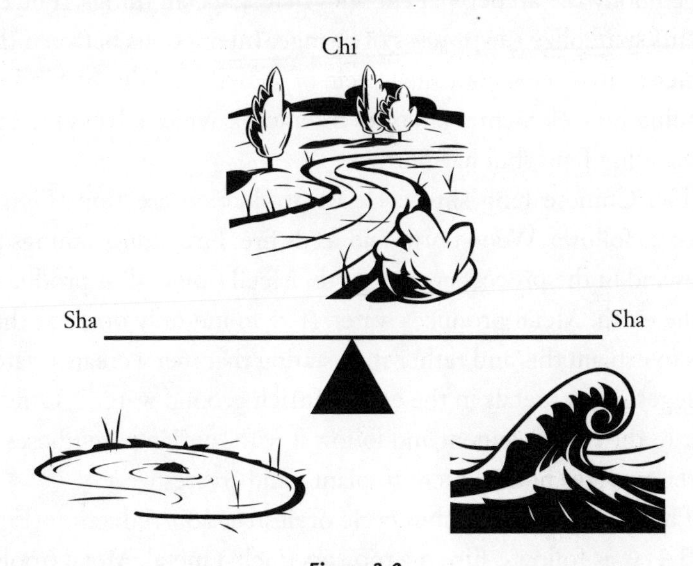

Figure 3.2
Balanced chi is healthy energy. Extreme yin or yang results in unhealthy energy (sha).

can feel overwhelming. In most cases this can be caused by an overabundance of one or two elements that can be brought under control.

The important thing is to be aware of how energy is moving (or not) in and around your home. You may need to slow fast-moving energy, get it activated where it accumulates in a corner, or invite more of it into your home. Through various means you can affect the quality of the energy, which in turn will have an effect on you. The art of feng shui is itself a balancing act of protection and creation: protection from negative energy and creation (or enhancement) of healthy energy. This use of energy allows you to change your life.

The Five Traditional Feng Shui Elements

According to the ancient Chinese, all things in the universe can be divided into five elements: water, fire, wood, metal, and earth. These five elements are utilized in traditional Chinese feng shui and embody the archetypal energies that shape all things. The elements symbolize the process of change. Interactions between the elements produce a continual cycle of growth and decay. Understanding how elemental energies interact provides a basis for understanding feng shui harmony.

The Chinese feng shui cycle of production/creation (Figure 3.3) is as follows: Wood burns and feeds fire. Fire reduces things to ashes and in the process creates earth. Metal (mineral) is produced by the earth. Metal produces water. (I've found only one text that tries to explain this and rather than stating that metal creates water it suggests that metals in the earth "enrich ground water."[4] Others refer to the metal element and follow it with "air" in parentheses.[5]) Water provides nourishment to plants, and creates wood.

The traditional feng shui cycle of destruction/reduction (Figure 3.4) is as follows: Fire overpowers (melts) metal. Metal (tools) cut and destroy wood. Wood consumes earth by draining nourishment. Earth overpowers water (by directing the flow of rivers and streams). Water extinguishes fire.

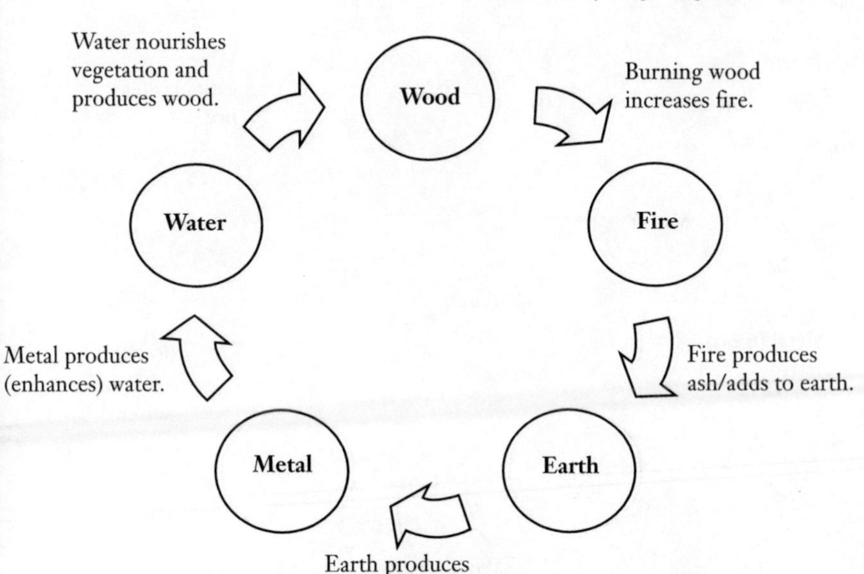

Figure 3.3
The feng shui elemental energy cycle of production/increase.

Just as the four Western elements mentioned in chapter 2 have corresponding attributes, the Chinese feng shui elements also have attributes such as directions and seasons with which they are associated. Table 3.2 contains a few basic correspondences. Refer to a traditional feng shui text for the full range of associations.

In Chinese, the five elements are called *Wu Xing: Wu* is "five" and *Xing* means "to move." In a nutshell, this is the significance of the elements: cycles of movement and change. You might want to try the seasonal/elemental meditation explained in chapter 2 again, but this time using the traditional feng shui colors and directions. If you noted your first experience in a journal, check for differences that you may have experienced using these elements.

When working with elemental energy, try to include all elements to provide balance. However, if an element is too strong in a particular room, its effect can be balanced with another element

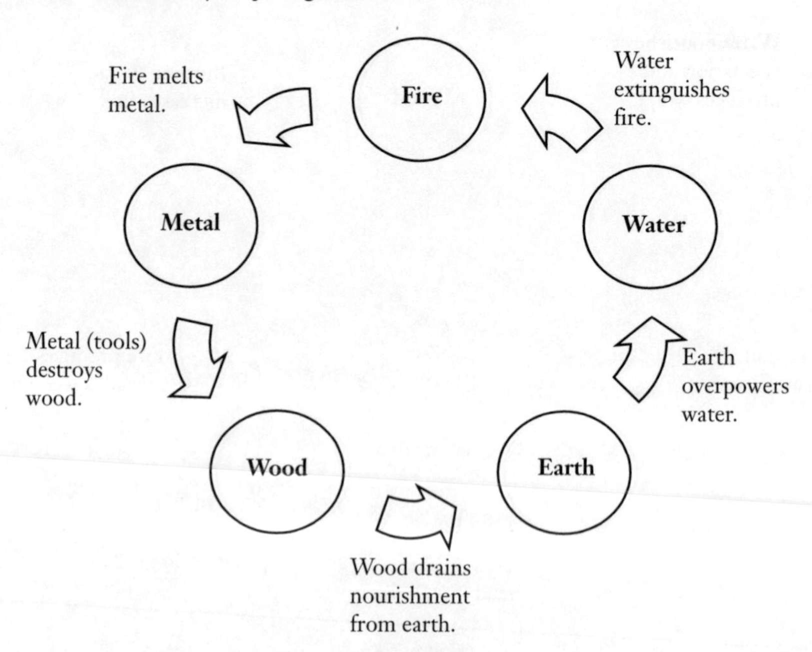

Figure 3.4
The feng shui elemental energy cycle of destruction/decrease.

Table 3.2—Traditional Feng Shui Elemental Attributes

Element	Season(s)	Direction(s)	Basic Color(s)
Fire	Summer	South	Red
Metal	Autumn	West and northwest	White and gray
Wood	Spring	East and southeast	Green and purple
Earth	Late summer and late winter	Southwest and northeast	Yellow and blue
Water	Winter	North	Black

to bring the room into harmony. If you use one element to tone down another, it is important to be aware of any disharmony that may be created by using an element's destructive quality. For example, if water is utilized to decrease the effects of fire, it has the potential to completely remove fire so that it is no longer present. The intention itself (annihilation) can create disharmony in the room. It is preferable to use an element of moderation, which will reduce the power of another element without destroying it (Figure 3.5).

Balancing elemental energy can be accomplished by using the element itself or a representation of it. In gemstone feng shui, the color and other attributes of stones are used to represent an element. The size and number of stones as well as their placement can emphasize and increase the presence of an element.

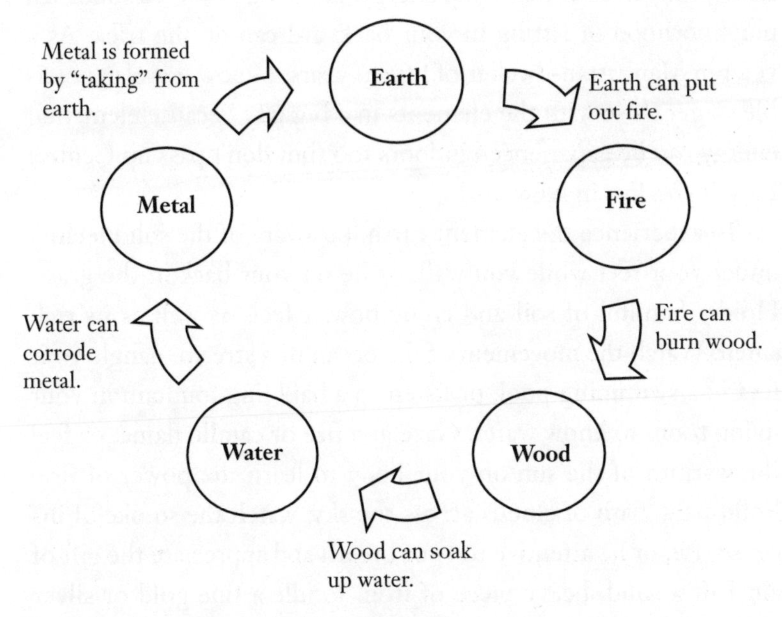

Figure 3.5
The feng shui elemental energy cycle of moderation/reduction.

Get in Touch with the Elements

The power of the elements and their associated directions are dynamic through the cycle of the year (seasonal, macrolevel) and individually (personal, microlevel). In addition to your personal element, you may be drawn to different elements and their associated directions and gemstones at different times throughout your life. Explore each of the elements to see how they may help you balance energy flow. It is important to be aware of the effect an element has on you. Experiment with different elements and intensity of elemental energy to find what best suits your individual energy.

I find that the best way to be in touch with the elements is to spend time out of doors. A hike in the woods is ideal for feeling connected, however, you don't have to wander too far from your door to experience elemental energy. Strolling around a suburban neighborhood or sitting in your backyard can do the trick. As a veteran Manhattan-dweller of fifteen years, I know it is also possible to get close with the elements in a big city because elemental energy can be experienced indoors too (but don't pass up Central Park if you live in New York).

To experience the element earth, be aware of the solid feeling under your feet while you walk or lie on your back in the grass. Hold a handful of soil and enjoy how it feels as well as its rich smell. Watch the movement of the ocean or a stream, dangle your feet in a swimming pool, or listen to a babbling fountain in your living room to know water. Gaze at a fire or candle flame, or feel the warmth of the sun on your body to learn the power of fire. Follow the path of clouds across the sky, watch the smoke of incense rise, or be attentive to your breath and appreciate the gift of air. Lift a solid, heavy piece of iron, fondle a fine gold or silver chain, or catch the glint of pyrite mixed with another mineral to experience metal. In all instances, focus on the various qualities of each element and the sensations they bring you (Table 3.3).

Table 3.3—Elemental Flow Direction and Associations

Metal
Energy flow: inward
Associations: abundance, wealth, success

Fire
Energy flow: ascending
Associations: momentum, action, transformation, the intellect

Earth
Energy flow: horizontal
Associations: stability, reliability, confidence

Water
Energy flow: descending
Associations: the emotions and the ability to go with the flow

Wood
Energy flow: outward
Associations: growth and creativity

The Landscape: Body of Mother Earth

Not only has modern culture taken people away from the land, it has also been instrumental in destroying the land. When roads or housing estates are built, the land is leveled, which allows energy to rush past. Without the natural contours to provide give and take, the health of moving energy has been traded for fast highways and suburban sprawl which are eventually replaced with more of the same.

In contrast, the ancient people of Europe left their legacy in stone across the Continent, Mediterranean islands, and British Isles in the form of circles, alignments, single-standing stones, and dolmens (chambers formed of standing stones). Several thousand structures were built between 5000 and 500 B.C.E. Many theories abound as to how and why they were built, but many agree that some of the sites accurately track the rising and setting of both sun and moon at the winter and summer solstices. The placement of these ancient monuments in relation to hills and mountains suggest that the landscape itself functioned as part of these structures.

The use of sacred geometry (mathematically determining the relationship of an object with its surroundings to maintain a harmonious balance) has been noted by people studying the placement of ancient monuments. The landscape seems to envelop these sites except for one side, which has an open view of the horizon. This type of harmony with the earth seems to have been used in later times for the placement of the palace of Knossos in Crete, and the temple at Delphi, the spiritual center of ancient Greece. According to Rachel Pollack, the Earth was seen as the Great Mother Goddess and it was important for a temple or sacred site to be enveloped and protected by the Great Mother's body.[6]

In addition to their immediate surroundings, the sacred monuments of Europe and Egypt, as well as those in the Americas and Easter Island, are aligned on energy channels called *ley lines.* These lines form a web around the entire Earth and act as an energy grid. The intersection of ley lines creates power nodes where a great deal of energy accumulates. The ancient Chinese geomancers referred to them as *lung mei* or "dragon paths," and noted polarized energy lines as either green dragon (yang) or white tiger (yin).[7] The energy itself that flowed over and through the land was called "dragon's blood." These channels are not unlike the energy meridians in humans, which are manipulated in acupressure and acupuncture to stimulate chi and good health.

In today's landscape flattened by progress, it is not easy to position a home in an optimal, earth-embracing location. However, the surrounding buildings and trees may help recreate this flow of energy to provide protection and comfort.

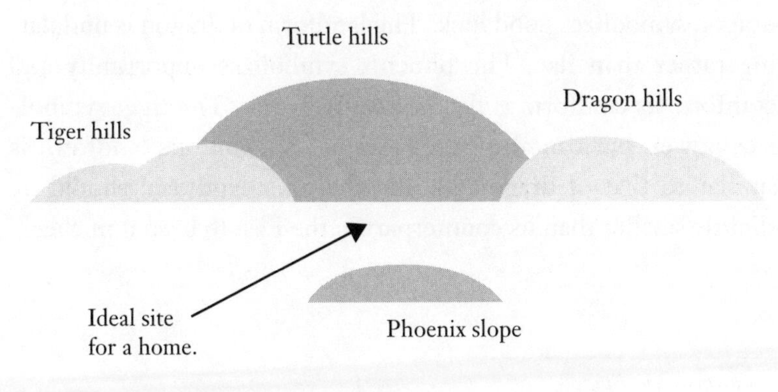

Figure 3.6
The ideal setting for a home.

Ancient Chinese Landform Energy

Like the ancient Europeans, the Chinese also saw the importance of location and landscape. Fitting into and living in harmony with the natural environment is a powerful way to ensure healthy energy in your home. As in European temples, the ideal location for a home should provide support in back with a mountain or high hills, gently envelop both sides with smaller hills, and provide an open view in front. If there is a gently flowing stream in front, so much the better. Jami Lin has likened this setup to an armchair with a high back, protective armrests, and a footstool in front, and symbolizes a comfortable life.[8] Four celestial guardian animals were equated with this ideal configuration: black turtle, green dragon, white tiger, and red bird (phoenix) (Figure 3.6).[9]

In the best of all worlds the ideal configuration would also align the guardian animals with their respective directions: turtle in the North, dragon in the East, phoenix in the South, and tiger in the West. Turtle symbolizes strength and support. The landform would resemble a turtle's back. It is psychically and physically reassuring to have your back supported. In China, the green dragon symbolizes good luck. The landform of dragon is undulating rather than flat. The phoenix symbolizes opportunity and comfort. Its landform is flat, or relatively flat. The tiger symbolizes power, but can also be dangerous, so while its landform is similar to that of dragon, ideally the tiger landform should be slightly smaller than its counterpart in the East to keep it in check.

---

1. Govert, *Feng Shui*, 8.
2. Too, *Basic Feng Shui*, 8–10.
3. Starhawk, *The Spiral Dance*, 32.
4. Wu Xing, *The Feng Shui Workbook*, 20.
5. Lin, *Contemporary Earth Design*, 81.
6. Pollack, *The Body of the Goddess*, 17.
7. Lin, *Contemporary Earth Design*, 278.
8. Ibid., 146.
9. Too, *The Fundamentals of Feng Shui*, 12.

CHAPTER FOUR

Introduction to Tools

The Bagua

There are various schools of feng shui that are based on the use of compass directions, landforms, or symbols. Some disciplines utilize a combination of devices to read and work with energy. The first of these is the *bagua.*

The *I Ching* was written to help people deal with change in their lives, as well as to get in tune with the cycles of nature. It has been said that feng shui is the *I Ching* manifest. The eight trigrams of the *I Ching* are believed to be universal symbols through which life can be understood. These symbols consist of parallel lines where a broken line represents yin and an unbroken one yang.

The trigrams indicate long-term patterns of change and are associated with various seasons, family members, and natural phenomena (Table 4.1). While the trigrams are assigned to certain seasons (or parts of seasons), the actual yin/yang combinations of their lines are associated with particular members of the family and their auspicious directions on the Magic Square. For example, the most yang of line combinations (chien) represents father and the most yin (k'un) represents mother.

The octagonal bagua (depending on the school of feng shui, it is also referred to as the *pa kua*) contains the eight trigrams and is used for feng shui analysis as well as protection. Hung on the outside of

Table 4.1—Trigram Correspondences

Trigram	Name	Direction	Season	Family Member
☵	K'an	North	Winter	Middle son
☶	Ken	Northeast	Late winter	Youngest son
☳	Chen	East	Early spring	Eldest son
☴	Sun	Southeast	Late spring	Eldest daughter
☲	Li	South	Early summer	Middle daughter
☷	K'un	Southwest	Summer	Mother
☱	Tui	West	Autumn	Youngest daughter
☰	Chien	Northwest	Late fall/early winter	Father

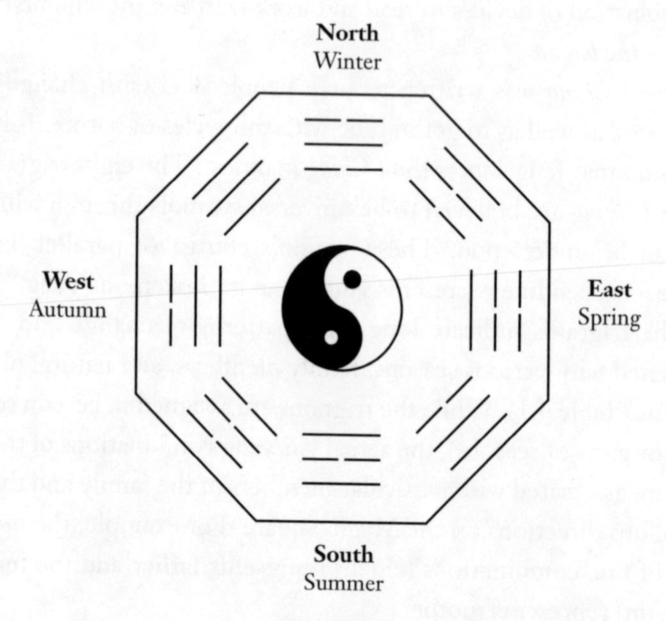

Figure 4.1
The bagua using the Latter Heaven Arrangement. In this illustration, North is shown at the top. The traditional feng shui bagua would show South at the top.

the home, the bagua deflects negative energy. The organization of the eight trigrams where they are positioned to represent the cycle of the year is referred to as the "Latter Heaven Arrangement" (Figure 4.1). The "Former Heaven Sequence" represents a perfect universe.[1]

Like the traditional bagua, a bagua of gemstones utilizes the power of Earth energies and represents the turning of her seasons. The quarter days of the year—winter solstice, spring equinox, summer solstice, autumn equinox—and their associated gemstones provide a foundation for invoking the essence of the seasons. The annual cycle of the year provides seasonal balance of yin and yang (yin/winter and yang/summer). The days of equinoxes furnish equal hours of day and night, light and dark, hence, equal yin and yang.

In his exhaustive study of the *I Ching*, feng shui master Lin Yun, adjunct professor at San Diego State University, analyzed the various meanings of the book's title. He interpreted it as meaning "changeable/unchangeable" and consisting of two characters: one representing the sun, the other the moon.[2] Daily, these celestial bodies provide a balance of light and dark—yin and yang—which have their own associated gemstones. A few of the gemstones linked with the sun include red stones such as ruby, garnet, and spinel. These stones represent the fiery brilliance of our local star. Moonstone and pearl are associated with the moon. Their white luminosity is evocative of Luna's cool comfort.

A bagua of gemstones can be as large or small as you care to create. Begin by gathering stones. If you want to create a bagua that represents the seasons, gather a piece of black spinel, jade, red garnet, and white quartz. If you don't have these particular stones, you can use ones that are the appropriate colors: black, green, red, and white. Find a flat surface such as a table or windowsill upon which to place the stones wherever you sense the need to balance energy. Use a magnetic compass to determine which direction is North. Imagine a circle and place the black stone in the North part

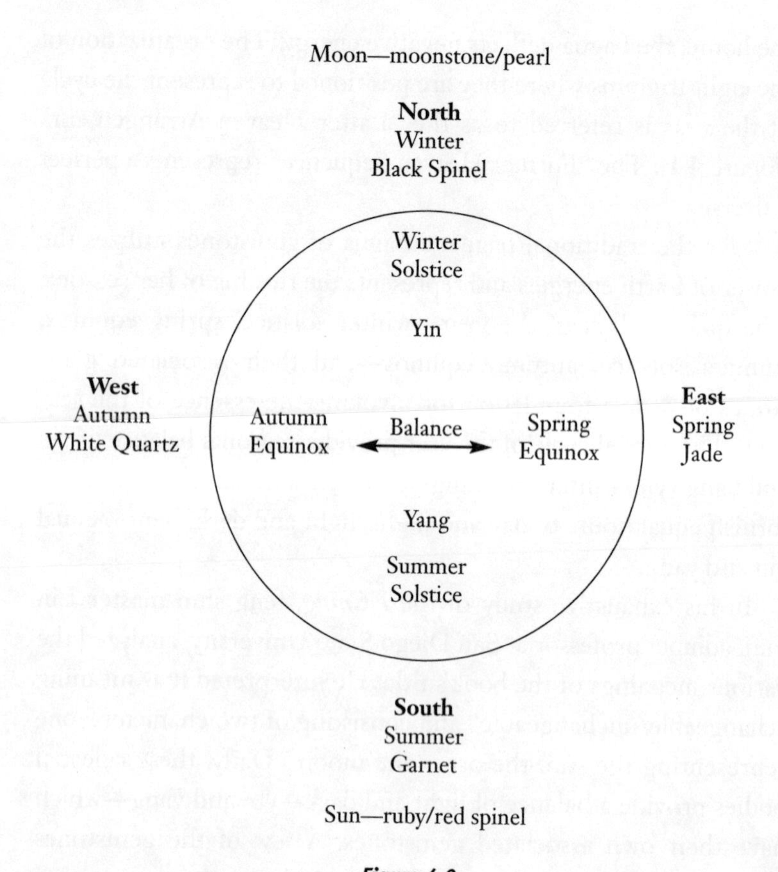

Figure 4.2

A bagua of gemstones to represent the seasons. Representing the moon and sun is optional.

of the circle. Place the red stone opposite on the South part of the circle. Place the green and white stones opposite each other in the East and West positions, respectively. If space is limited, use small stones and place them close together (Figure 4.2).

If you are artistically inclined, you could draw or paint the outline of a bagua on a piece of posterboard and glue the stones into position. This could be propped up on a table or hung on a wall.

For more of a challenge, create a three-dimensional mobile by suspending the stones with string from the posterboard bagua.

Creating a bagua of seasonal gemstones provides balance and a reminder of the cycles of the year (and life). This helps to keep us in touch with the natural world and to keep our energy grounded. As you read further in this book you will find other types of baguas to create (using the nine feng shui directions or five elements). They are all constructed in the same way by placing gemstones in a circle or square according to their associated directions.

The Pa Tzu *Compasses*

In the aptly named Compass School of feng shui, the basic reference tool is a compass. In the past, the *lou pan* (compass) of feng shui masters was a complicated instrument with eight to thirty concentric rings of code words that referred to the trigrams, elements, *lo shu* numbers, and other attributes used in feng shui analysis. This is also referred to as the Chinese geomancer's compass.

At the center of the lou pan is a compass that would be more familiar to us today except that it referenced South rather than magnetic North to which we are accustomed. A simplified version of the lou pan is the *pa tzu compass* containing a trigram, element, lo shu number, and energy group identification (Figure 4.3).

The nine numbers on the compass are used to determine the direction and element that holds the most power for you. After calculating your number, find it on the compass. (In traditional feng shui this is referred to as the *lo shu*, or *kua, number.*) The calculation is based on your year of birth and is different for men and women. The center of the compass holds the direction number five. If after calculating the number for your power direction your result is five, use number two if you are a man, or number eight if you are a woman (Table 4.2).

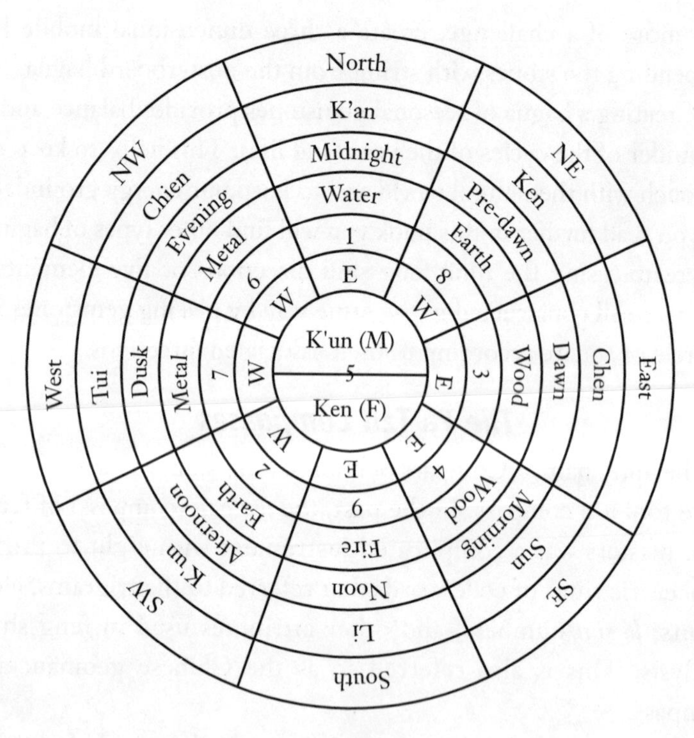

Figure 4.3

The feng shui pa tzu compass. A traditional compass would show South at the top.

Table 4.2—Determining Your Lo Shu Number

Subtract the century
from your birth year.
(E.g., 1947 – 1900 = 47)

For Women	**For Men**
Add 5 to the result from above. (E.g., 47 + 5 = 52.)	Add the digits until you get a single number. (E.g., 4 + 7 = 11, and then 1 + 1 = 2.)
Add the digits until you get a single number. (E.g., 5 + 2 = 7.)	Subtract from 10. (E.g., 10 – 2 = 8.)
If your result is 5, use the number 8 as your lo shu number.	If your result is 5, use the number 2 as your lo shu number.

Wheel into Square

The numbers of the nine directions are derived from the Lo Shu Grid. This grid is based on the remarkable pattern found on the back of a tortoise that China's first emperor, Wu of Hsia, found in the River Lo.[3] The grid itself resembles our modern pound sign (#) as well as the *kanji* character for the *Ching*, or the Well, hexagram of the *I Ching*.[4]

The Well hexagram represents a method for drawing upon the deep strength within the mystery of life. In Western civilization the "Great Mysteries" were celebrated by Neolithic people on the island of Malta. There, people scratched the curved walls of Hal Saflieni Hypogeum out of solid rock to create an underground chamber in which to practice their sacred rites.[5] In ancient Greece, the nine-day rites of Demeter and the Eleusinian Mysteries acknowledged with awe and respect the divine spark from which we were created and which is within the deepest part (the well) of our souls. The Well hexagram is connected with feng shui in that it is a symbol of reaching deep inside oneself to connect with one's own energy, which is necessary in order to touch the energy (and essence) of life.

After extensive analysis, ancient Chinese scholars believed that the markings on the tortoise provided a perfect three-by-three magic square (the Lo Shu Grid or River Lo Map) (Figure 4.4).

6	1	8
7	5	3
2	9	4

Figure 4.4
The Lo Shu Grid.

6—Northwest	1—North	8—Northeast
Achievement/ Benefactors Assisting People Father	Career Personal Journey Foundation	Wisdom Knowledge Self-cultivation
7—West	**5—Center**	**3—East**
Creativity Projects Children	Balance Harmony	Family Ancestors Community
2—Southwest	**9—South**	**4—Southeast**
Relationships/ Partnerships Love/Romance/ Marriage Mother	Illumination Fame/Success Reputation/ Respect	Prosperity/Blessings Abundance Self-/Net worth Resources

Figure 4.5
The Magic Square.

The numbers, if added in any direction, equal fifteen. This is also the total number of days in a complete waxing or waning cycle of the moon, which includes the dark and full phases of Luna. Awareness of the moon and her cycles complements the use of Earth energy and adds to our ability to balance yin and yang in our lives.

The Lo Shu Grid, combined with the compass, produces a matrix of attributes and associations for each of the directions. This matrix is called the *Magic Square* (Figure 4.5).

East—Includes all those you love from your immediate family to the family of humankind, your ancestors, and friends. This area is also about growth and vitality.

Southeast—Wealth and abundance on all levels from self-worth to monetary worth and blessings received. This area also includes personal resources and anything that enriches your life. Go here to recover your self-worth.

South—To be known in your community or field of employment, or garner respect and recognition. This is concerned with your outward self as well as self-actualization. If you want to shine, work on this area.

Southwest—Encompasses love, romance, marriage, and all types of relationships—personal and business. Many times we learn about relationships through our mothers. This direction is also associated with being receptive.

West—If you want to have children, help your children, or stimulate your creative impulses, activate this area. This is where you nurture all manner of things to which you "give birth."

Northwest—Achievement also includes those who help you achieve: benefactors, mentors, and your father. It is about responsibility—giving, as well as receiving.

North—Careers and personal journeys symbolize progress in our lives. Progress is not always about moving forward; at times we must stand still and assess ourselves before we can move onward. Think of this as your area of foundation.

Northeast—Wisdom, knowledge, and self-cultivation also encompass turning points. Gaining self-wisdom can bring about revolutions in our lives.

Center—This is a place of balance, harmony, and spirituality; also considered the prosperity point. To attain balance and harmony requires joy.

To use the Magic Square, you will need a magnetic compass. Once you determine which way is North, orient the Magic Square to it. Imagine superimposing the Magic Square over a room in your house. It may be easier to draw a floor plan of the room, then draw the Magic Square on a piece of tracing paper and lay it over

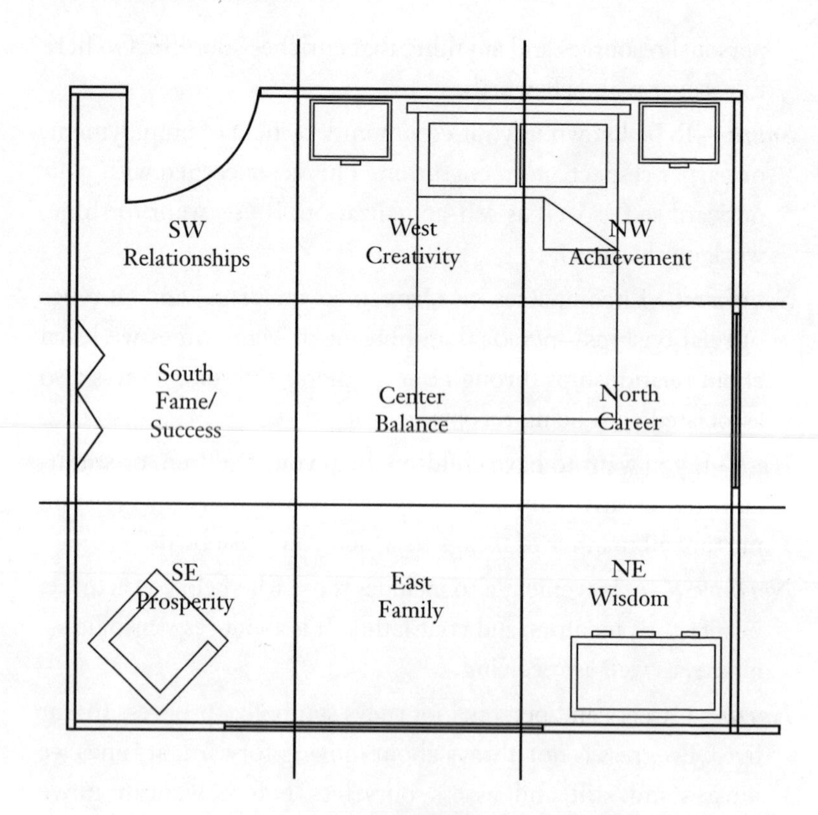

Figure 4.6
The Magic Square superimposed over a room.

the room sketch. This divides the room into nine sectors, each corresponding to a direction and personal aspects (Figure 4.6).

If you are experiencing a particular problem in your life, identify what sector it would fall into. For example, if you are having problems in your career (or if you want to grow and expand into a different career), assess what is in the North sector of the room. Take a few minutes to sit in this sector. Close your eyes and try to sense what is happening with the energy there. If the area is cluttered or has something such as a piece of large furniture protruding into it, the energy is not flowing smoothly. Get rid of any clutter and relocate anything that might be impeding energy flow. To encourage a healthy flow of energy, place an associated gemstone in that area of the room.

If you are in the midst of a major career transition, you may want to emphasize and honor the North direction. You could create a small separate area with crystals and gemstones, or you may simply want to place a few crystals or stones on a shelf or table in this area. Use whatever feels right for you to honor this direction and coax the energy to flow freely. Alternatively, you might also check that there is not an abundance of wood energy in this area—wood drains nourishment from earth in the elemental cycle of destruction. This might be impeding the flow of earth energy.

By applying the directions, elements, and aspects of the Magic Square to your life, you are able to activate and enhance the energy around you and augment your personal energy and connections to the natural world. It can help you with healing or sorting out a particular situation. Before you rearrange a room or your entire house, take the time to determine what you want or what you want to move toward.

What aspect on the Magic Square do you feel you need to encourage or work with in some way? Figuring out what you want to accomplish ahead of time will make the process of feng shui easier and more effective.

Widening Your Scope

Just as the Magic Square is used to assess a room, it is also utilized to assess an entire house or apartment (Figure 4.7). The ideal position of your house will have its back to your power direction. For example, if you have calculated your lo shu number on the compass as one, your power direction is North. The best position for your house is with its back to the North (also called "seated" in that direction) with the front facing South. If it is not seated in your power direction, the next best place would be one of your power points.

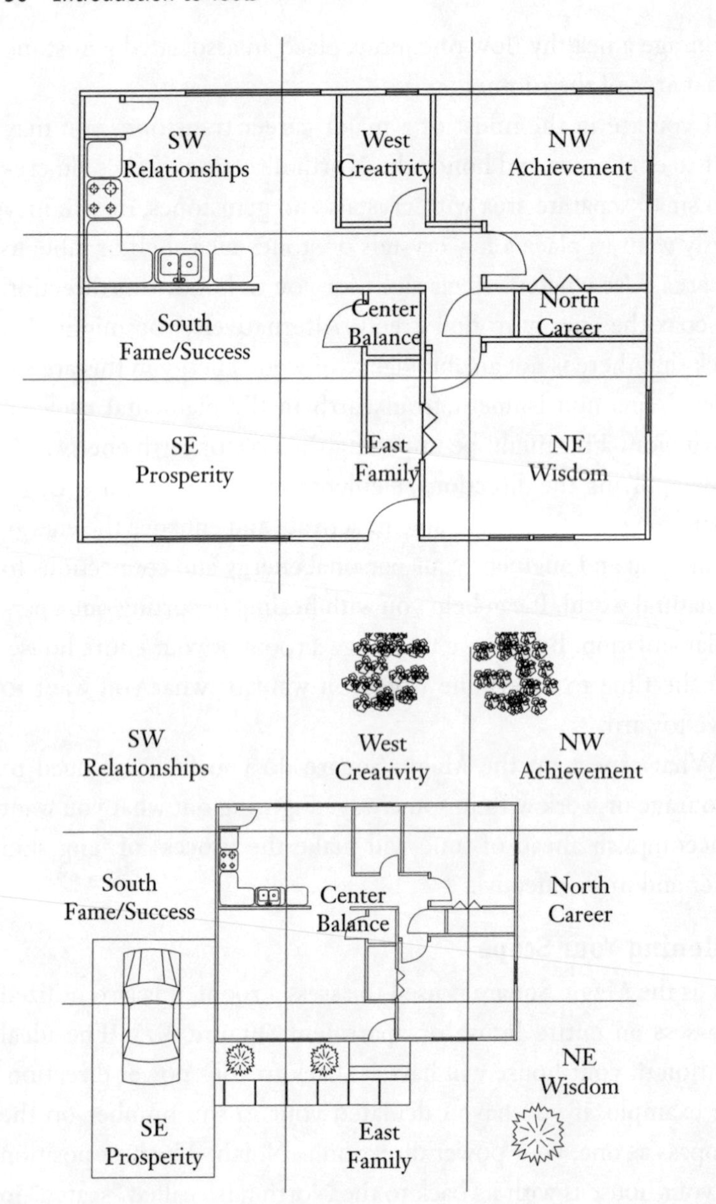

Figure 4.7

The Magic Square superimposed over the floor plan of an entire house and property.

Table 4.3—Energy Groups and Power Points

West Energy Group		East Energy Group	
West	Metal	East	Wood
Northwest	Metal	Southeast	Wood
Northeast	Earth	South	Fire
Southwest	Earth	North	Water

Energy Groups and Power Points

Each direction is part of an *energy group* that is indicated on the pa tzu compass with an *E* or *W*. Each energy group contains four directions. Find your power direction in one of the energy groups; the other directions in that group are your *power points* (auspicious directions) (Table 4.3).

In addition to your personal element mapped on the compass, the elements of your energy group also have an influence on you. For example, if your direction is Southeast, your personal element is wood, but fire and water also have an effect on you. The West energy group has only two elements, and a person of this group is generally influenced strongly by both.

Having your home seated in a direction that gives you power is important because it is the base from where you draw strength and support. It allows you to draw on the natural energies that are most appropriate for you. For example, when you are sitting you are probably most comfortable with your back against something solid because it provides you with support and security. Likewise, an unpleasant situation is easier to handle with people you trust behind you, or in other words, backing you up.

If your home is not seated in one of these auspicious directions, you can use a variety of objects to emphasize your power direction throughout your home to help you draw on the energy you need for balance. Emphasize your power direction and power points by making sure that energy can flow freely in the areas that correspond to these directions. Increase the strength of the elements associated with these directions in their respective areas by using gemstones that evoke each element directly (such as peridot for fire in the South) or gemstones that represent the color of a direction (such as moonstone for gray in the Northwest). Refer to Figure 6.1 on page 71 and Figure 7.1 on page 83 respectively for details.

You might also want to assess the energy of the property on which your house or apartment building sits. To do this, use the following visualization exercise. This can also be used to assess your house or one room: Imagine a slow, gentle wave of shallow water entering the front of your property. Think of how the water would move as it encounters objects. Is there anything that would inhibit it from smoothly continuing on its path? Are there areas where it might stop moving? Is this something that would create a funnel where the water would be forced into a narrow course, which would cause it to move faster? Energy moves throughout your home and property in the same way. Most problems of energy flow are not difficult to remedy and are addressed in subsequent chapters.

In each situation, note the areas of your power direction and power points and be sure to keep healthy energy moving in them. While you might be most comfortable with two or three particular elements, it is important to utilize all five in order to achieve overall yin/yang balance. In this way the energy in your home will be complete.

Positive and Negative Directions

While your energy group indicates the directions that give you power, the opposite energy group indicates the directions that bring you challenges. In traditional feng shui the positive directions (the ones in your energy group) are called *Prime*, *Health*,

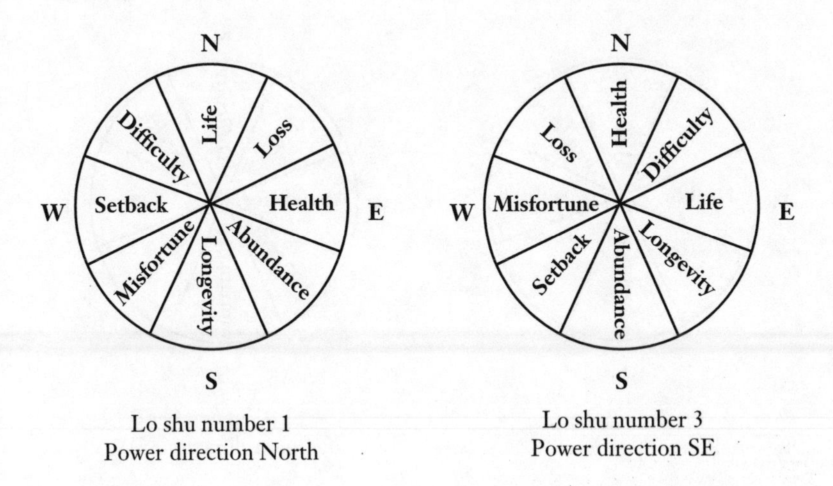

Lo shu number 1
Power direction North

Lo shu number 3
Power direction SE

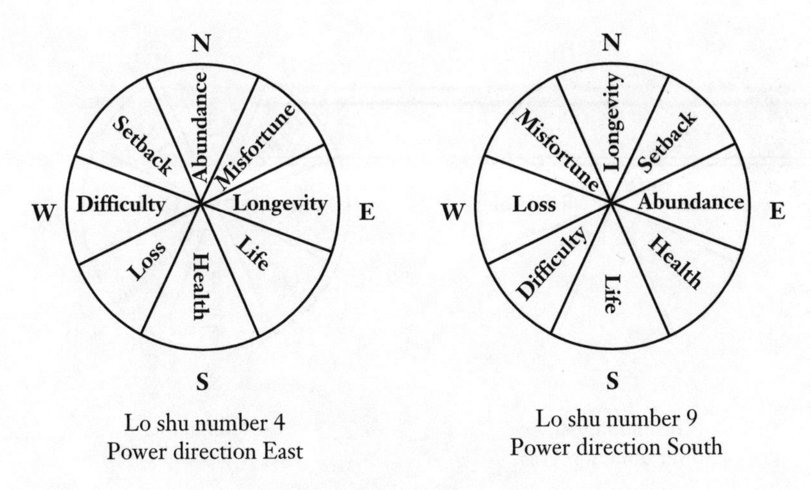

Lo shu number 4
Power direction East

Lo shu number 9
Power direction South

Figure 4.8
East energy group's positive and negative directions.

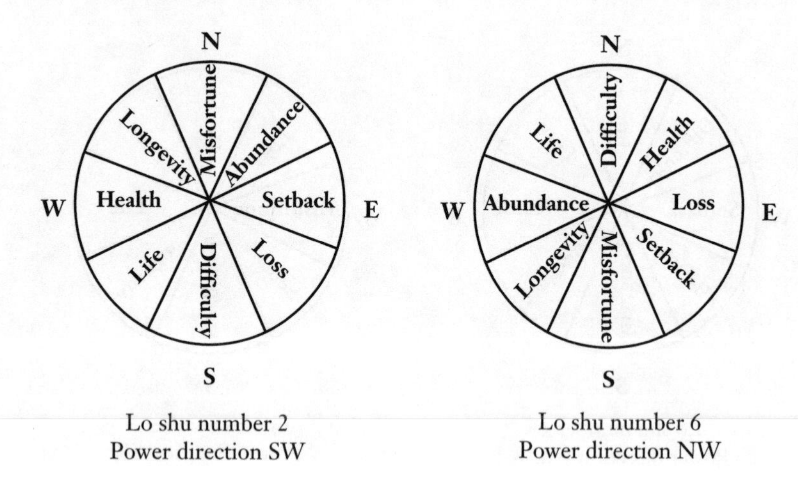

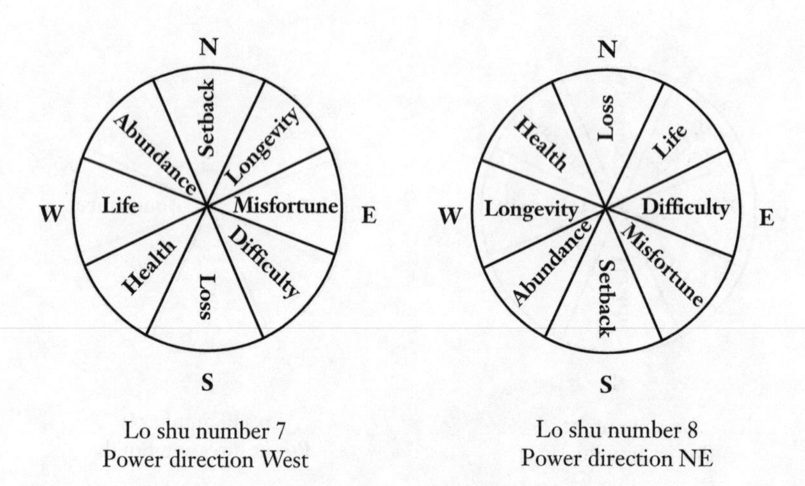

Figure 4.9
West energy group's positive and negative directions.

Longevity, and *Prosperity*. The negative directions (the ones in the opposite energy group) are called *Five Ghosts, Six Shar, Disaster,* and *Death*. The various schools of feng shui use slightly different names. They are also called *Life, Health, Longevity, Abundance, Misfortune, Setback, Difficulty,* and *Loss*.

Whatever they are named, the negative directions do not indicate the kind of impending doom that some of their names suggest. Rather, these should be viewed as areas that present particular weaknesses or issues. A propensity for energy that is negative for you accumulates in these directions, which in turn effects any weakness or issue. Your specific positive and negative directions are determined by a combination of your compass number and energy group (Figures 4.8 and 4.9).

Life—This is concerned with abundance and prosperity in family and career; things that constitute "the good life." This aspect occurs in your power direction and is the ideal direction in which your home to be seated.

Health—This area is connected with vitality and good health; a zest for life. If you are ill or low on energy, this is a good area to activate. It is a good location for a kitchen or dining room.

Longevity—As its name implies, this area has to do with long life and, of course, good health. Not only is this a good place for your kitchen or dining room, but also the master bedroom.

Abundance—This area is concerned with well-being and prosperity. It is a good place for the front door, the kitchen, study, or work area. Activities that affect the well-being and prosperity of the household should be performed in this area.

Misfortune—This area has to do with accidents or illness and a general lack of energy or vitality. It is a good location for the bathroom or laundry room so that misfortune can be flushed or washed away. This is not a good place for the front door as you would be inviting misfortune into your life.

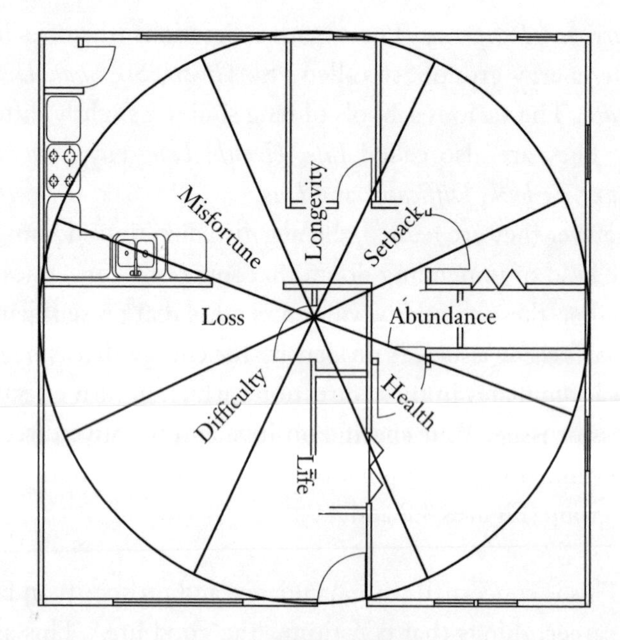

Figure 4.10
A wheel of positive and negative directions superimposed over a floor plan.

Setback—This area is connected with unexpected events and disputes. This is a good place for the bathroom, or storage closets and other types of areas that are not frequently used.

Difficulty—This area is concerned with disappointments and annoyances. It is a good area to use for disagreeable chores such as ironing, and for allowing yourself to let go of things that bother you.

Loss—This area is connected with the loss of prosperity and property through events such as theft or financial problems. This is a good location for a bathroom, or storage closets and other types of areas that are not frequently used.

Like the Magic Square, your energy wheel can be superimposed over your floor plan to determine where these areas fall in your home. In the example in Figure 4.10, the front door is in the Life area, which essentially invites "the good life" into the home. The Misfortune area in the kitchen presents a problem as this could invite illness and a lack of energy into a room that is vital to a family's health.

Methods for countering negativity depend on the cause of the problem. For example, remedy stagnation produced by blocked energy flow with something that moves such as wind chimes or a fish aquarium. The cutting effects of sharp energy can be softened with plants, fabrics, or crystals. More information for counteracting negative energy can be found throughout the next chapter. Also refer to Table 7.3 on page 91.

Feng Shui with Family and Roommates

Your personal space—bedroom, study (or simply desk area), workshop, or sewing room—should be balanced for your energy. Others in your household will need to carry out feng shui for their own areas. You can do this for small children to ensure a good healthy atmosphere for them. In common rooms it is important to balance energy so it is favorable for all.

Because the living room is usually a place of activity, using feng shui in this room can have a strong effect on the energy of the entire house. In addition to the Magic Square aspects, areas of the living room (family room and house) relate to specific members of the household. These are based on the eight trigrams of the *I Ching* (refer to Table 4.1 on page 40). Taking into consideration today's blended and extended families, you may need to adjust the area assignments to correspond with those in your household.

In the bedroom, consider your power direction and power points for positioning the head of the bed (and your crown chakra). These will provide you with vital energy for renewing your body as you sleep. If your partner's energy group is the opposite of yours, you will need to work out a position that is comfortable for both of you. You may want to try positioning the bed where there are adjoining sectors such as Northwest and North, North and Northeast, Northeast and East, or South and Southwest. These adjoining sectors provide common ground and may furnish the key to a solution for you. Experiment to find what works best for you and your partner.

---

1. Webster, *Feng Shui for Beginners*, 45–46.
2. Lin, *Contemporary Earth Design*, 40, 47.
3. Webster, *Feng Shui for Beginners*, 77.
4. Govert, *Feng Shui*, 9.
5. Streep, *Sanctuaries of the Goddess*, 84.

CHAPTER FIVE

Traditional Use of Gemstones and Crystals in Feng Shui

Getting Started

There are three ways in which crystals and gemstones can be utilized in feng shui. One is using gemstones to invoke elemental balance in order to provide a healthy environment for yourself and your family. The second utilizes gemstone color associated with the Magic Square or Lo Shu Grid to change your life and aid in healing. The third method invokes the power of birthstones. Before learning how to employ these gemstone methods (which are covered in later chapters), it is important to understand the basics of traditional Chinese feng shui and how gemstones and crystals originally figured into it. Suggestions for the further use of gemstones and crystals have also been included here.

When beginning feng shui (traditional Chinese or gemstone), it is important to take your time and assess your environment

Table 5.1—The Three As of Creating Good Feng Shui

1	Awareness	Assess your surroundings for sources of negative energy
2	Adjustment	Correct and protect against negative energy
3	Activation	Create and enhance positive energy flow

before jumping in and changing things. Table 5.1 shows the three basic steps, or the three *A*s, of creating good feng shui.

Performing feng shui on a room or an entire house (or apartment) does not mean that you have to change your complete decor, rearrange all your furniture, or bring in a bulldozer to clear away accumulated stuff. In many cases it is as simple as rearranging a few things or introducing a new object or two into a room.

The key to success is to start small. Don't try to take on the whole house at once. Begin with one room or one area within a room. Take your time, change things mindfully, then take time to live with it to assess how it affects you. If one thing doesn't work, try something else until you find what is right for you. What works for a friend or relative may not work for you because everyone's personal energy is unique. Don't be afraid to try something completely different if your heart of hearts tells you it's right for you.

It is important to maintain awareness of the good feng shui you create because this is a dynamic process. Energy flow changes with the weather and the seasons just as your own energy may change. Most people feel more bright and cheerful on a sunny spring day than on a wet winter evening. Our personal dynamics change and so do our homes.

Poison or Secret Arrows

Poison arrows are sharp angles that focus a harmful point of energy. Poison arrows can be projected by the corner of an interior wall, sharp-cornered furniture, or the edge of a neighbor's roof that points toward your house. The first step is to identify any potential poison arrow inside or outside your home.

In Figure 5.1, the poison arrow from the protruding wall could cause problems—tension, headaches, upset stomach—to a person sitting on the sofa across the room. One way to remedy this is to stop the poison arrow by placing an object (such as a round table) in front of the sharp angle. Another way is to hang a small crystal from the ceiling approximately six inches from the angle. This

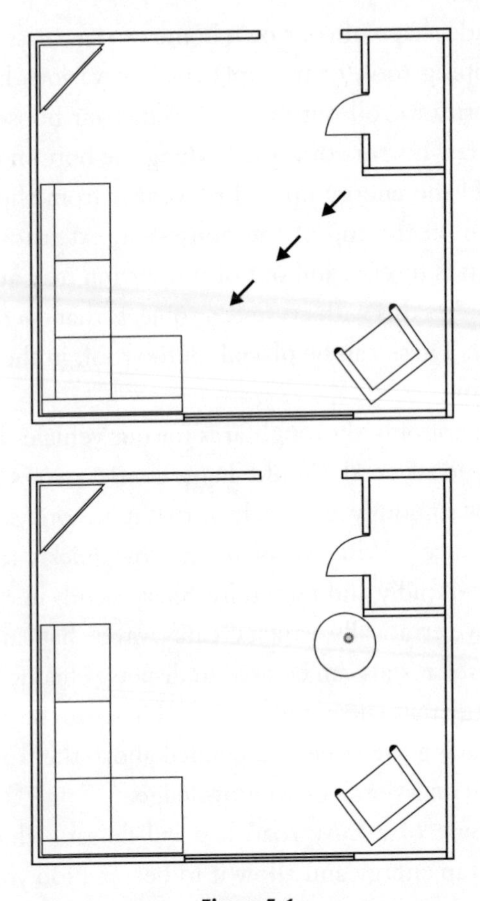

Figure 5.1

A poison arrow from the sharp angle of a protruding wall can be remedied with a round table in front of the corner.

neutralizes the sharp energy with positive crystal energy. Arranging a gemstone bagua on the table would also neutralize the negativity. If the angle of a neighbor's roof or other structure outside your home produces a secret arrow, a crystal or gemstone on a windowsill or other area where the energy strikes your home will disperse the negativity. Protective crystals and stones include agate, aquamarine, carnelian, lapis lazuli, malachite, and obsidian.

The outside shape of your own home can cause issues for you. A steeply sloping roof, or multiple roofs at various heights, can cause prosperity to roll out or away from your house. Rounded gutters, or a few mirrors or crystals along the bottom of the roof, will help hold the energy up and prevent it from slipping away. Any structure on the top of the house—an extension, chimney, antenna—that is too tall and out of proportion may attract negativity. Counteract this with crystals or objects that encourage beneficial energy. These can be placed on the roof, in the attic, or in the room below.

Roads are not only thoroughfares for our vehicles but also for energy. The ideal road should have gentle curves that allow energy to flow smoothly and freely so that it is more yang than yin but is still balanced. Wide roads that narrow quickly force energy to move more rapidly and forcefully. Sharp bends in a road cause energy to move erratically, which creates waves that can crash into nearby houses. Crystals can be used to disperse chaotic energy and calm the energy that enters your home. These can be arranged on windowsills facing the street, suspended above the front door, or placed outside on a porch or window ledge.

The flip side of a busy road is a cul-de-sac, which has the potential to trap energy and allow it to become too yin. Crystals, gemstones, and other feng shui devices can be utilized to activate energy. As long as there is some movement and stagnation does not occur, a cul-de-sac can actually create a pool of calm energy with just the right amount of yang for balance.

The Entrance to Your Home

Your front door provides the main access for energy to enter your house. If anything causes a blockage here, energy will not enter and circulate freely to create a balanced environment. It is important that energy has a clear, even route of access to the main entrance of your home. It also needs space (and time) to begin the transition from outer world to inner. The energy that you need to affect the aspects of your life is about to cross your threshold and you can usher in positive rather than negative things. This is where you set intention about yourself and about what you want; your entrance is a message to the world about who you are. If the approach to the world of your home is gentle and friendly, the energy you need will come to you.

The front door and porch should be in proportion to the house itself. An entrance that is small will allow in an amount of energy that is appropriate for the doorway, but perhaps not enough to circulate all the way through the house. Blessings and abundance could be stuck on your doorstep. Likewise, an entry that is sized too grand for a house will invite in more energy than is appropriate and overwhelm the space within. Crystals can help regulate the flow of energy.

If your home is situated opposite a T-junction roadway, fast-moving energy is bombarding your front door. If you live in an apartment and your front door is at the end of a long corridor, energy is rushing down the hall and overwhelming your home. A crystal hung near the outside of your front door will provide protection against the rushing stream of energy as well as smooth the flow before it enters your home.

A front door that opens opposite a wall in a small foyer can inhibit energy from entering and circulating through your home. The energy that comes through the door encounters the wall and most of it will flow back out the door. Placing a mirror opposite the door would bounce the energy out even faster. To entice the energy in, hang a crystal above the door.

If your home has a long hallway that comes to a dead end (and no back door), it is important to keep the area clear and well lit. Energy can get trapped at the end of such a hall, so the use of crystals will help maintain circulation. If your home has a center hall with the front and back doors in alignment, energy can roll straight through before it has a chance to reach other rooms. A crystal near each door (and any other doorways along the corridor) will act as traffic cops to slow and circulate the energy.

Inside Your Home

The Dining Room

The dining room is a place for good eating and good socializing. It's a place where families come together not only for nourishment, but also quite frequently for family discussions or decision-making. A formal dining room may have a crystal chandelier in it, which adds movement and color to the room when the light strikes it. Any light fixture over a table—provided it is not too bright—will help stimulate energy. Crystals and gemstones on side tables will add beneficial energy to the room. Tall china cupboards in the dining room can produce sharp angles of energy like a protruding corner of a wall. The same remedy as before can be applied by suspending a small crystal or gemstone from the ceiling approximately six inches in front of the secret arrow.

Windows

Large windows allow more energy in than windows with small panes. Very large windows and sliding glass doors can create problems by allowing an overwhelming amount of energy to enter a room. A bowl of crystals or gemstones on the windowsill or a table in front of the window can calm the energy flow. Use stones that are smooth and round for this softening effect.

Fireplace

A fireplace is usually a welcomed feature in a home. In the past, the fireplace was the heart of the house. It was used for cooking and served as the main, if not the only, source of heat. Family members gathered before it and guests were accorded a seat close by to warm themselves. Anyone who has lingered in front of a crackling fire knows its magic and soothing ambiance. A fireplace and chimney provides a special gateway for energy to enter your home. Keep the fireplace clean. Gemstones and crystals placed around the hearth provide protection and help smooth energy entering your house.

Furniture

A couch is usually the most-used piece of furniture in a house and needs the support of a wall behind it. If your couch does not have this support and you feel it might be causing problems, position a sofa table behind it to dampen the energy. A crystal (or two) on the table behind the couch can provide a buffer of protective energy.

Avoid positioning a couch—either front or back—opposite an entrance. A couch facing a door leaves its occupants vulnerable to a rush of energy that may enter the room. This can leave you feeling drained and tired. Sitting with your back to a door leaves you vulnerable and unprotected. The natural flow of energy—through your house or through you—is from front to back. Think of standing in the ocean with your back to the breakers. Most likely you would feel vulnerable because you cannot see what is coming or use your hands to protect or brace yourself. With energy washing over you from back to front, you are not able to protect yourself from negativity. Personal energy radiates from the heart center. To avoid vulnerability—if placing the couch opposite the door works best in your living room—place a crystal above or near the door to cleanse the entering energy of any negativity it may be carrying.

Furniture pushed flat against the wall creates corners where energy can stagnate. To remedy problems that may arise in this situation, place a plant in the corner created by the furniture or hang a crystal above it. Other energy-activating gemstones such as citrine, garnet, jasper, or sardonyx can be placed on tabletops.

Exposed Beams

Exposed beams are great decorative features, but they can be a source of problems. Beams with sharp corners create secret arrows that point downward on a room's occupants. Because energy must flow around and under them, sitting beneath exposed beams creates a roll of energy over your head. If the best position for your couch or chairs is below a beam, you might want to encourage a climbing or trailing plant to grow along the beam and then hide a few gemstones in the leaves.

The Living Room

The living room has a special area called the *wealth corner.* This is in addition to attributes on the Magic Square that may fall into this area. The wealth corner is not determined by compass direction, but by its location in respect to the main entrance of your home. Stand in the doorway to your living room and take one normal-sized step into the room. Turn to the left and look up to the ceiling. This is your wealth corner. If your living room has two or more entrances, use the one that is closest to the main entrance to your home (Figure 5.2).

A small crystal suspended from the ceiling about six inches from the wealth corner will activate beneficial energy. In traditional feng shui, a small stack of shiny coins, a green object, or a picture of something that symbolizes prosperity is also used. In gemstone feng shui, crystals and stones particularly associated with attracting prosperity include beryl, chrysoprase, iolite, and moonstone.

Angled and Irregularly Shaped Rooms

Occasionally when a large old house has been split into apartments, the result is some unusual room shapes. In recent years, architects have taken to providing uniqueness to new homes by constructing angled or irregularly shaped rooms. Many times what gets created is a general feeling of imbalance and incompleteness.

Lay the Magic Square over this type of floor plan and you will find a piece missing (Figure 5.3). If this happens to be your power direction or one of your power points, a piece of you is missing from the room. The lack of energy flow in such an area can be compensated for by placing a mirror or crystal on the angled wall.

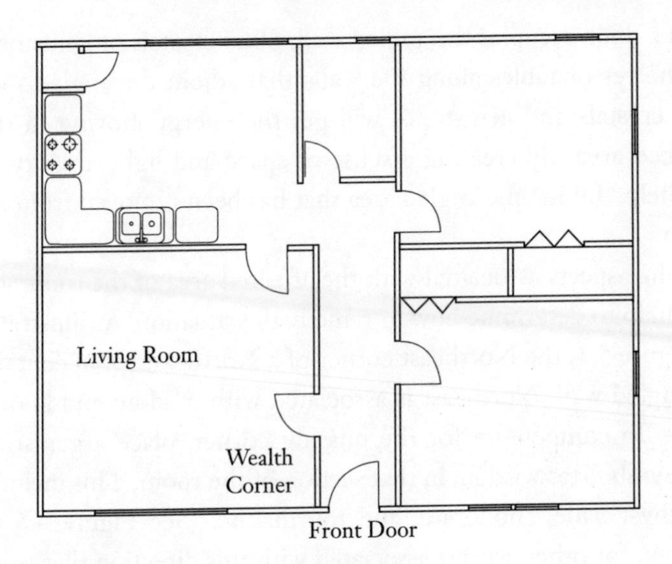

Figure 5.2

The wealth corner.

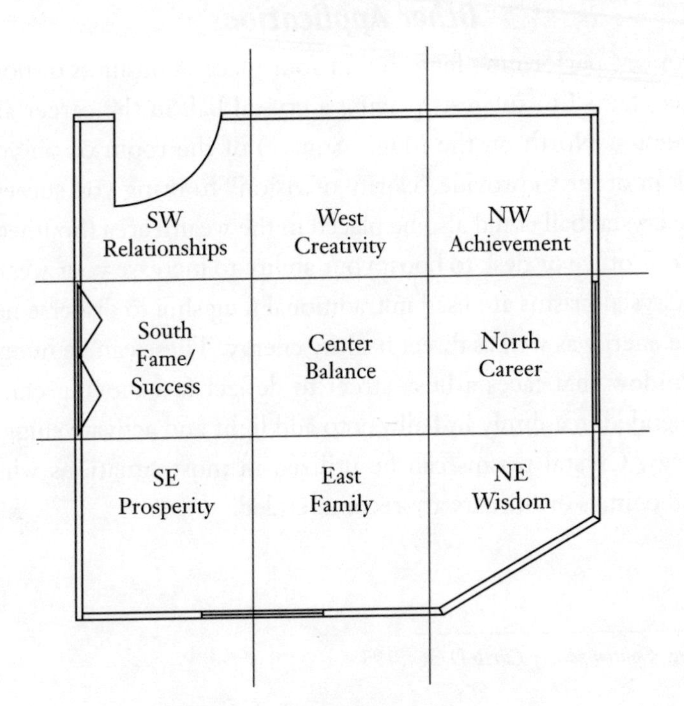

Figure 5.3

Angled rooms cut off certain aspects of your life.

If a door occupies the angled wall, place crystals or gemstones on shelves or tables along the walls that adjoin the angled wall. The crystals and gemstones will get the energy moving in this blocked area. By creating a sense of space and light, the crystal will help "fill in" the angled area that has been removed from the room.

The aspects associated with the affected area of the room will also help to determine how to remedy the situation. As illustrated in Figure 5.3, the Northeast corner of a Northeast room contains an angled wall. Northeast is associated with wisdom and knowledge. To compensate for the missing corner, place a gemstone that symbolizes wisdom in that section of the room. This includes amethyst, jade, rhodonite, and tourmaline. (See Figure 4.5 on page 46 for other aspects associated with this direction that could be affected.)

Other Applications

If you are performing feng shui in your place of business or home office, Jami Lin suggests placing a crystal ball in the career area (direction North on the Magic Square) of the room or on your desk in order to provide "clarity of vision" to bring you success.[1] The crystal ball could also be placed in the wealth area (Southeast) of your office or desk to boost your ability to increase your wealth.

Crystal prisms are used in traditional feng shui to disperse negative energy as well as direct healthy energy. These can be hung in a window that faces a busy street to deflect fast-moving chi, or suspended in a dimly lit hallway to add light and activate sluggish energy. Crystal prisms can be utilized in most situations where wind chimes or mirrors are recommended.

---

1. Lin, *Contemporary Earth Design*, 195.

CHAPTER
SIX

Gemstones for Elemental Balance

Natural Energy Flow

As was mentioned in the previous chapter, one method of using gemstones and crystals in feng shui is for elemental balance. When energy enters your house it needs to circulate freely in order to maintain the vitality of its life force and yours. Blockages will cause energy to stagnate and become too yin.

As was also previously mentioned, it is best to start small when applying feng shui principles to your home to avoid being overwhelmed by the task. However, it is important to develop an overview of your entire home before making changes. A problem with energy flow in one room may have its root cause in another or in the general movement of energy through the house (upstairs and down if you have multiple floors).

One room can affect another. A room can have all the right colors, objects, and placement of things for good balance, however, if the room next to it is emanating negativity, the first room will be affected. A room where the elements are following the cycle of destruction will drain the energy from others. When the elements are out of kilter, extreme yin or yang will throw off the balance of the entire house.

Begin with an assessment of the overall floor plan and halls that connect the rooms, then assess each room individually.

Determine the areas that require adjustment and prioritize their importance. Tackle one at a time and reassess your overall plan when you make changes. Don't immediately assume that you will have to make changes. Many parts of your home probably already exist in harmony and balance.

In some rooms you may want to emphasize a particular element, or there may be elements that you are more comfortable with than others. You will most likely find that the elements associated with your power direction and energy group will be more welcoming for you. While the other elements may not provide a warm and fuzzy feeling for you, it is essential to have all five elements represented in your home for overall energy balance and health.

When the elements are in balance, the harmonious energy created is called *sheng chi*, also known as the "dragon's cosmic breath." This intrinsic energy of perfect balance invites good fortune into our lives.

Certain gemstones are associated with the five feng shui elements. The primary stones or their alternates can be utilized (Table 6.1). You can also refer to an elemental Magic Square for information on how to balance energy and/or affect certain aspects of your life (Figure 6.1).

A quick glance at Figure 6.1 reveals a telling picture. The most prominent element is earth, which occupies three of the nine sectors. We can think of ourselves as being born from Mother Earth and we return to her when we die, so a greater proportion of earth helps keep us grounded. Wood and metal have two sectors each and help to control fire and water (one sector each), which are strong elements with great destructive powers. This grid illustrates the basic proportions of elements for balance and what area of a room or house will possess more of a particular element.

Create a scale drawing of the floor plan of your apartment or each floor of your house. Make a Magic Square to the same scale on a piece of tracing paper. With a magnetic compass, determine

Table 6.1—Element/Gemstone Associations

Element	Direction	Gemstone	Alternate
Fire	South	Peridot (born of fire; found in volcanic rocks and meteors)	Obsidian (formed from hot lava)
Metal	West Northwest	Malachite (57 percent copper)	Azurite (another copper-related mineral)
Wood	East Southeast	Jet (formed from woody plants)	Amber (from ancient tree resin)
Earth	Southwest Northeast	Andalusite (known as the "earth stone")	Tourmaline (called "something little from the earth")
Water	North	Opal (can contain 3–10 percent water)	Pearl (formed by sea creatures)

Northwest Metal Malachite Achievement	**North** Water Opal Career	**Northeast** Earth Andalusite Wisdom
West Metal Malachite Creativity	**Center** Earth Andalusite Harmony	**East** Wood Jet Family
Southwest Earth Andalusite Relationships	**South** Fire Peridot Fame/Success	**Southeast** Wood Jet Prosperity/Worth

Figure 6.1
Elemental Magic Square.

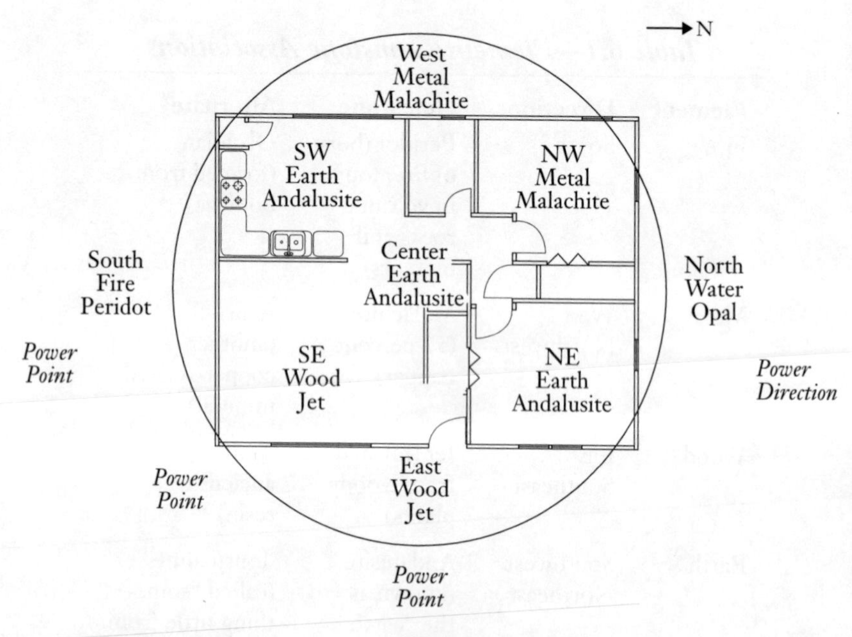

Figure 6.2
Directional elements superimposed over a floor plan. This example illustrates the power direction and power points for a person with the pa tzu compass number one and who is in the East energy group.

which direction is North. Orient your floor plan in the correct direction, then lay the Magic Square over it. This orientation of the Magic Square will remain the same for each individual room (Figure 6.2).

Fire

For many people, the kitchen is the heart of the house. It is where friends and family gather to share food, conversation, and life. It is a place that engenders warmth and security. Elemental balance and the movement of energy through the kitchen and dining area are important since cooking and eating are vital to health, healing, and well-being.

Traditionally, South and Southeast are considered the best directions for a kitchen; the Southeasterly winds were considered auspicious for lighting a fire. In the cycle of elemental production, wood is burned to create fire. In the illustration in Figure 6.2, part of the kitchen occupies the South sector of the floor plan. More importantly, in the kitchen itself, the stove is positioned along the South wall and can draw on the strength of purifying fire.

It is also important to be sure that fire does not overwhelm the other elements in the kitchen or cause a "struggle." A sink or refrigerator positioned next to the stove can cause elemental disharmony because these represent the element water, and water extinguishes fire. If you have this setup in your kitchen, position pieces of jet and andalusite between the stove and sink or refrigerator to moderate both elements. Andalusite represents earth, which moderates fire, and jet represents wood, which moderates water. Align the stones so that the jet touches the sink but not the stove (wood increases fire). Allow the andalusite to touch the stove.

Place peridot in the South sector of your home or room (wherever you are practicing feng shui) to attract success. The size of the gemstone that you employ for feng shui is not as important as the intent you use when situating the stone in a particular place. Clarity of intention is essential to manifest desires into the physical world.

The weather and seasons affect the energy of a room and the potency of an element. On dreary, rainy days we feel down because of the strength of the darkness, damp, and chill, or yin energy. On such a day, bring up the yang energy in your home by emphasizing fire energy. Place peridot in areas that need a lift. Since the strength of fire is diminished in the cold, in winter months you may want to keep extra peridot and other fire-invoking gemstones to place around your home. Then, as fire is strongest in the summer, you would put the extra stones away.

The fire element is also amplified during the full moon, which would be a good time to cleanse your extra fire stones before storing them.

As its name suggests, the living room is for living and should have more yang than yin energy. The living room is usually the center of relaxation and social activity. Because it is most often the largest and one of the busiest rooms in the house, the extra yang energy created here feeds vitality to the rest of your home. If you have a fireplace in your living room, den, or family room (all areas of family-oriented activity), a location on a South wall increases the yang energy of fire. During the summer months when the fireplace is not in use, you may want to place a small piece of peridot on the mantel to maintain the level of yang to which you are accustomed. Because the fire element is stronger in the warmer months, it won't need much of a boost.

Remember the elemental cycles of increase and decrease. The use of jet in the South will help to increase fire, as jet represents wood, which builds fire. Opal—which represents water—placed in the South would subdue the effects of fire, as water extinguishes fire. If you find that you need to tone down the fire element, use a little bit of andalusite, which represents earth, in the South. Earth controls fire but does not destroy it.

Water

Water is the other strong element that holds only one position on the Magic Square. Its direction is North.

Generally, the water element is useful near the entrance of your house to help energy (and abundance and prosperity along with it) flow into your home. It also helps to calm and smooth out fast-moving energy before it enters. If you have a small fountain or birdbath in your front yard, hide a piece of opal near its base to amplify the water element. If you don't have a water feature there, or if you live in an apartment building, suspend a piece of opal above the front door or place it on the door frame above the entrance.

While giving the water element a boost in the front of your house helps good things flow into your life, strongly emphasizing water behind your house does the opposite and drains prosperity away. If you have a stream, pond, or other body of water directly behind your house, you may want to moderate the water's draining effects by placing pieces of jet on the windowsills that face the water. Also, remember that water destroys fire, so use opal carefully and with clear intention in the South sector of your home or property. Be in tune with the overall balance of elements.

Inside the home, water requires tricky balancing. The bathroom is a place where unneeded and unwanted material and energy is removed from the body. It is important that negativity not continue from this room into others. On the other hand, the bathroom is also a place for cleansing and reuniting (very strongly) with the water element. By nature, this is a difficult room to bring into balance. The major problem encountered with the bathroom is that energy moving through the home is easily flushed or drained away here.

Ideally the bathroom should be situated at the back of the house—energy moves from front to back—which helps to reduce the possibility of spreading negativity to other rooms. Counter the water element with jet (wood moderates water). Where the water element feels extremely strong and uncomfortable, you might employ the use of a small piece of andalusite. However, because earth destroys water, you will want to use this with a clear intention of not totally obliterating the water element. Getting rid of an element completely will not allow the other elements to exist in balance. If you have difficulty balancing the elements in the bathroom, the traditional solution is to place a mirror on the outside of the door. This will symbolically make the room disappear. A clear or white quartz crystal will achieve the same result.

The least favorable location for a bathroom is in the Center of the home. A quick check of the Magic Square will show you that the Center is the place of spirit, balance, and harmony for the entire house. Because it is also thought of as the heart of the house, the draining action of water here can pull the soul out of your home. Definitely use quartz in this situation.

Another place where you might want to make the bathroom's negative energies disappear is if it is located where it is the first thing seen when someone enters your house. Draining and elimination is not the first impression you want to portray for your home. In addition to keeping the door closed, place a quartz crystal and a piece of azurite or amazonite just outside the room. These gemstones will transform the energy and provide it with more yang life force, as well as help prevent negativity from greeting your guests and family when they enter your home.

In the bedroom, an en suite bathroom is a convenience, but it can drain the energy away (Figure 6.3). A healthy energy flow in the bedroom is vitally important to the quality of your sleep, and in turn, your general well-being. The best solution is to keep the bathroom door closed, especially if it is located directly opposite the bedroom door or windows, to prevent energy from being swept directly away. Place gemstones such as amethyst or jet on the bedroom windowsill to calm and slow the energy so it doesn't immediately drain toward the bathroom. A few other gemstones placed around the bedroom will encourage the energy to circulate.

If the only place to position your bed is opposite the doorway to the bedroom or the en suite bathroom, use protective gemstones such as ametrine, coral, or tourmaline to provide a healthy flow of energy around you when you sleep.

Earth

The earth element is vital for our overall well-being. Since ancient times, people have revered the land as Mother Earth. Scientists

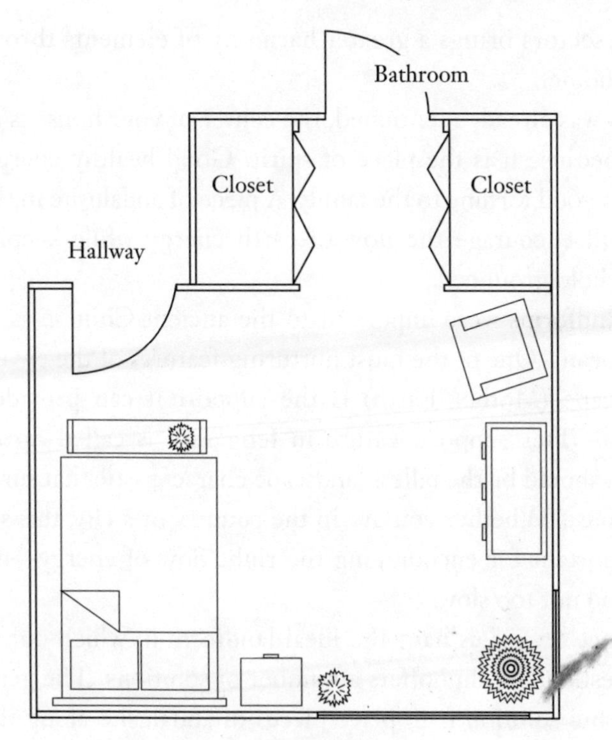

Figure 6.3
The en suite bathroom can drain the energy from your bedroom. Also, the door from the hallway here is a potential poison arrow.

have put forth the Gaia theory, which states that the Earth is a single organism of which we are a part. In feng shui, earth is the element at the core of the Magic Square—it is the orbit and movement of the Earth that ushers in the change of seasons and brings us night and day. It is the Earth's cycles that we follow. As individuals, we feel most healthy and strong when we are "grounded."

Because earth is so important to us, this element occupies three sectors on the Magic Square—Southwest, Center, and Northeast (Figure 6.1, page 71)—to form an axis around which the other elements are arranged. Strengthening earth energy in these sectors brings a greater harmony of elements throughout your home.

As was already mentioned, the center of your house is important because it is the place of spirit. Good healthy energy here brings good fortune to the family. A piece of andalusite in this sector will encourage the flow of earth energy while keeping the household grounded.

Landforms were important to the ancient Chinese as well as Europeans. One of the most nurturing features of the enveloping landscape (Mother Earth) is the support it can provide from behind. This support feature in feng shui is called *turtle hills*, which should be the tallest landscape characteristic that surrounds the house. Whether you live in the country or a city, this support is important for encouraging the right flow of energy—not too fast and not too slow.

Since few of us have the ideal landform in which our homes can nestle, feng shui offers a number of solutions. The gemstone feng shui solution is to place pieces of andalusite along the rear perimeter of your property. These can be positioned along or attached to a fence, or included in a decorative garden arrangement. This can also be combined with one of the traditional solutions of planting trees in the back of your property. Andalusite can be added to a birdhouse or feeder, or suspended from trees in any way that seems right to you. If you live in an apartment, place the andalusite on rear windowsills or along a back wall. Begin by using a couple of the gemstones to see how the energy feels to you, and add more if necessary.

The earth element is useful in grounding sharp energy whether it is produced from a protruding corner of a wall or furniture, or outside from an angled roof or other sharp structure pointing toward your house.

Wood

Wood symbolizes growth and renewal. As the element of the East, it represents spring and morning. Because wood is usually part of a living organism, its energy is very yang. While it is important to have yang energy circulating freely throughout your home, the bedroom is a place where you would want to de-emphasize it.

A healthy energy flow in the bedroom is vitally important to the quality of your sleep, and in turn, your general well-being. Attention should be focused first on the master bedroom, or the bedroom where the breadwinner(s) sleeps, since their health and ability to support the household is important. Unlike other rooms where you want to encourage yang energy, the bedroom is a place that should have slightly more yin in order for the room to be conducive to rest. This can be a fine line to tread as too much yin can tip the scales of balance and create a tomb-like or depressing atmosphere. Be sure there is some yang energy present. If you use jet or amber in the bedroom, make it a small piece.

Unlike an adult's bedroom, children's rooms need more yang energy to strengthen and support their growing minds and bodies. Children need to absorb a great deal of vital life-force energy to keep healthy. A couple of pieces of jet can provide the stimulation necessary for smooth-flowing yang energy. However, do not overdo it. Do not use jet close to the bed. If you find your child is having trouble getting to sleep or has a problem studying in his or her room, check to see that yang energy hasn't been amplified to the point where it is chaotic and distracting.

As the element for the East and Southeast, wood amplified in these sectors strengthens their associated areas of life—family and community, and wealth and resources, respectively. Wood energy is outward-moving, which helps us reach out to other people. The use of jet can aid in strengthening relationships. Because wood encourages growth and expansion, it is good to utilize jet in a study area or any place you use for work or handling finances. A piece of jet on the Southeast corner of a desk can be helpful.

Metal

The metal element symbolizes prosperity. The energy of this element is inward-moving, which can aid in drawing prosperity and abundance to you. Place a piece of malachite in the wealth corner of your living room. Metal energy is also dense, so the results that you want to manifest with it may take time to accomplish.

The directions associated with metal are West (creativity and projects) and Northwest (achievement and benefactors). The associated life aspects are long-range or long-term, commensurate with the dense, slower-moving energy of metal.

Metal is useful for providing protection—think of a metal shield. Use malachite to disarm a sharp angle or poison arrow. If your home is located inside a fork in the road, cutting energies are being generated by the scissors-like roadways. Place pieces of malachite along the perimeter of your property that edge the roads to counteract their negative effects (Figure 6.4).

Be careful when using malachite in the East and Southeast sectors of a room (or area where you are performing feng shui), as metal destroys wood and can easily upset the balance if employed too strongly here.

Elemental Shapes

The elements not only have associated colors, seasons, and directions, but also shapes. These shapes are symbolic of the element's movement of energy (Table 6.2).

Whenever possible, purchase stones for elemental balancing that are close to their associated shapes. The combination of gemstone vibration and shape will enhance the strength and purity of its elemental energy. The wavy shape for water can be created

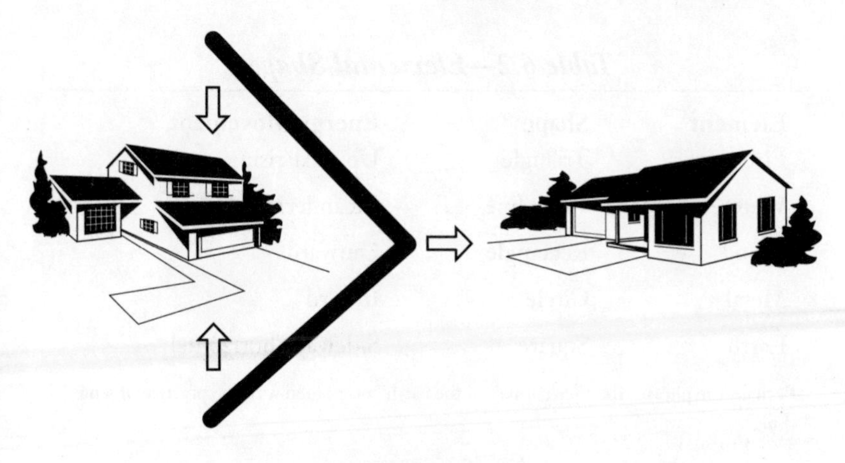

Figure 6.4
A forked road or a sharp bend in the road can create a cutting effect or poison arrow, depending on where your house is located.

with small pieces of opal laid out in a wavy pattern, or use the alternate gemstone—a string of pearls.

Another way to employ the use of gemstones for elemental balance in a room is to create a gemstone Magic Square. Ideally, this should be created in the center of the room. For this you will need three pieces of andalusite, two pieces each of jet and malachite, and one piece each of peridot and pearl. Try to have all the stones of a similar size. Since peridot is most commonly available in very small pieces, one or more of these gemstones may work better. Arrange all the stones in three rows according to their positions on the Magic Square. Be sure to orient your gemstone Magic Square to the appropriate directions. Leave this Magic Square in place for at least a day or two before assessing the effect it may have on the room's energy flow.

Remove other elemental objects from the area that needs balancing and set up the gemstone Magic Square. Once the area feels balanced, reintroduce the other objects one at a time to assess the individual effect on the energy of the room. You may decide to

Table 6.2—Elemental Shapes

Element	Shape	Energy Movement
Fire	Triangle	Upward/rising
Water	Wavy line	Meandering
Wood	Rectangle	Outward
Metal	Circle	Inward
Earth	Square*	Sideways/horizontal

*Perhaps the phrase "the four corners of the earth" originated with the practice of feng shui.

leave the gemstone Magic Square in place not only because of its effect, but also for its interesting arrangement.

If the gemstone Magic Square in the Center sector of a room works well, you may want to try it in the Center of your home. If the energy is balanced in the heart of your house, it will emanate harmony to other rooms.

CHAPTER SEVEN

Gemstones for Effecting Change

Color

Another method for employing gemstones in feng shui is to work with the colors associated with each direction and life aspect of the Magic Square (Figure 7.1). The elemental gemstones mentioned

Northwest—Gray	**North—Black**	**Northeast—Blue**
Moonstone Achievement Assisting People Father	*Black Spinel* Career Personal Journey Foundation	*Lapis Lazuli* Wisdom Knowledge Self-cultivation
West—White *White Quartz* Creativity Projects Children	**Center—Yellow** *Citrine* Balance Harmony	**East—Green** *Jade* Family Ancestors Community
Southwest—Pink *Rose Quartz* Relationships Love/Romance Mother	**South—Red** *Garnet* Illuminations Fame/Success Reputation/Respect	**Southeast—Purple** *Amethyst* Prosperity/Blessings Self-/Net worth Resources

Figure 7.1
Color Magic Square.

in the preceding chapter can be used in tandem with these stones to amplify to associated powers of both element and direction. Just as in traditional feng shui, utilizing or emphasizing these specific colors in their respective areas of a room can help influence an area of your life.

If you are experiencing a particular problem in your life, identify what sector it would fall into and then examine the sectors of your room, house, or property. To encourage a healthy flow of energy, place an associated gemstone in that sector of the room or the area where you are performing feng shui. For example, if you or your child are studying for an important exam, place a red garnet in the South sector of the room or the South sector of your desk to draw in success. You might also use a blue topaz in the Northeast sector as this is the place of knowledge and wisdom.

Feng shui is not only for correcting problems, but also achieving results. Decide what it is that you want to move toward in your life. If it is to build community, try using jade in the East sector of the room where you are performing feng shui. If you are working to advance your career, use black spinel in the North sector.

While each sector has a primary gemstone, any stone, as long as it has the corresponding color, can be utilized. Table 7.1 is a quick reference chart of gemstone colors. Where stones are known by different names, both have been included.

Healing

Also assigned to the sectors of the Magic Square are parts of the body that can benefit from the use of feng shui, or may suffer because of bad feng shui (Figure 7.2). Working on these areas in your home can be done in conjunction with crystal therapy on your body.

For healing purposes, utilize a gemstone of the appropriate directional color (Figure 7.1, page 83) along with one specific to the ailment (Table 7.2). For example, if you have a headache you may want to lie down in the Northwest sector of your room and hold a piece of moonstone in one hand and aquamarine or

Table 7.1—Gemstones Listed by Color

Color	Gemstones
Black	*black spinel*, augite, cassiterite, chalcedony, chlorite, coral, diamond, diopside, epidote, jade, jet, magnetite, melanite, obsidian, onyx, opal, pearl, smoky quartz, tourmaline
Blue	*lapis lazuli*, actinolite, agate, alexandrite, aquamarine, axinite, azurite, benitoite, beryl, bismuth, celestite, chalcedony, chrysoberyl, cobalt, cordierite, corundum, crocidolite, diamond, dumortierite, fluorite, hemimorphite, indicolite, iolite, jade, labradorite, lace agate, lazulite, opal, quartz, sapphire, smithsonite, sodalite, spectrolite, spinel, tanzanite, topaz, tourmaline, turquoise, zircon
Green	*jade*, agate, alexandrite, amazonite, andalusite, apatite, augite, autinite, aventurine, beryl, bismuth, bloodstone, calcite, chlorite, chrysoberyl, chrysoprase, clinozoisite, diopside, dioptase, emerald, enargite, epidote, fluorite, garnet, heliotrope, hemimorphite, jadeite, jasper, kunzite, malachite, nephrite, opal, peridot, prasiolite, quartz, sapphire, smithsonite, sphene, spinel, tanzanite, topaz, tourmaline, tsavorite, turquoise, zircon
Purple (violet)	*amethyst*, alexandrite, ametrine, apatite, axinite, chrysoberyl, fluorite, garnet, iolite, jade, kunzite, purpurite, quartz, sapphire, spinel, scapolite, smithsonite, spinel, sugilite, tanzanite, tourmaline, violane, zircon
Red	*garnet*, beryl, bismuth, calcite, carnelian, chalcedony, cinnabar, coral, cuprite, fluorite, jasper, opal, pyrope, rhodonite, ruby, sapphire, spinel, sunstone, thulite, topaz, tourmaline, wulfenite, zircon

Table 7.1 continued

Color	Gemstones
Yellow	*citrine*, amber, amblygonite, ametrine, andalusite, apatite, beryl, bismuth, brazilianite, chalcedony, chrysoberyl, danburite, diamond, enargite, fluorite, jade, jasper, orthoclase, phenacite, sapphire, scapolite, smithsonite, sphene, titanite, topaz, tourmaline, witherite, zircon
Pink	*rose quartz*, apatite, beryl, bismuth, coral, danburite, diamond, kunzite, lace agate, moonstone, phenacite, purpurite, rhodochrosite, sapphire, scapolite, smithsonite, spinel, topaz, tourmaline
White	*white quartz*, agate, albite, anorthite, apatite, aragonite, beryl, calcite, celestite, chalcedony, chlorite, coral, diamond, fluorite, jade, labradorite, lace agate, milky quartz, moonstone, onyx, opal, pearl, sapphire, spinel, tourmaline, ulexite, witherite, zircon
Gray	*moonstone*, agate, albite, anorthite, aragonite, calcite, cerussite, chalcedony, enargite, hematite, iolite, pearl, smoky quartz, tiger iron

Northwest Head	**North** Kidney Ears	**Northeast** Hands
West Mouth Bladder	**Center** Spleen Pancreas	**East** Liver Feet
Southwest Abdomen	**South** Heart Eyes	**Southeast** Hips

Figure 7.2
Healing Magic Square.

Table 7.2—Healing Gemstones

Ailment	Gemstone
Acne	Rose quartz
Adrenal glands	Black tourmaline
Allergic reaction	Bloodstone
Anemia	Hematite, garnet, bloodstone
Anxiety	Aventurine, amethyst, amazonite
Arthritis	Lapis lazuli, topaz
Asthma	Malachite
Backache	Amber, green tourmaline
Bacterial infection	Malachite
Bladder	Bloodstone
Blisters	Rose quartz
Broken bones	Green tourmaline, topaz
Bronchitis	Aquamarine, aventurine
Burns	Blue-lace agate
Bursitis	Amber, blue-lace agate
Circulation	Bloodstone, rose quartz
Colds/Sinus	Azurite, amethyst
Constipation	Ruby, smoky quartz, black tourmaline
Cough	Aquamarine, turquoise
Diarrhea	Black tourmaline, smoky quartz
Digestion	Amber, carnelian, citrine
Earache	Amber, amazonite, aquamarine
Eczema	Sapphire
Eye problems	Aquamarine, beryl, chalcedony, obsidian, topaz, jade, opal
Fevers	Opal, ruby

Table 7.2 continued

Ailment	Gemstone
Gall bladder	Carnelian, citrine, malachite, emerald, green tourmaline
Hay fever	Blue lace agate
Headaches	Amethyst, aquamarine, turquoise, blue tourmaline
Heart	Citrine, onyx, ruby
Heartburn	Peridot
Herpes	Jadeite, lapis lazuli
Hip pain	Jadeite
Hot flashes	Moonstone
Infections	Malachite
Insomnia	Amethyst, smoky quartz
Intestines	Brown tourmaline
Legs	Jadeite, spinel
Kidney	Citrine, jade, onyx
Liver	Citrine, jade, topaz, opal, peridot
Measles	Turquoise
Menstrual cramps	Carnelian, moonstone
Migraines	Aquamarine, turquoise, blue tourmaline
Morning sickness	Moonstone, red jasper, malachite
Mumps	Aquamarine, topaz
Muscle aches	Jadeite
Pancreas	Agate, alexandrite, opal
Rash	Blue lace agate
Sciatica	Green tourmaline, smoky quartz
Sinus problems	Azurite, blue lace agate
Sore throat	Blue lace agate

Table 7.2 continued

Ailment	Gemstone
Spleen	Alexandrite, opal, pearl
Sprains	Aventurine
Stomach	Aquamarine, jet, peridot
Varicose veins	Amber, bloodstone

turquoise in the other. Different gemstones may produce different results. A stone that you are familiar with may provide a quicker-acting cure. As with all aspects of feng shui (and crystal therapy), continually assess the effects on your personal energy.

Table 7.2 provides a quick reference to gemstones used in healing. While we are all healers and can work to strengthen and heal our own bodies, minds, and spirits with alternative and holistic approaches to medicine, it is important to balance our self-work with modern medical practice. Work with your doctor and use gemstones to support your treatments.

In addition to the Magic Square, other traditional tools function in gemstone feng shui. By replacing the trigrams, a gemstone compass is created from the pa tzu compass (Figure 7.3). The gemstone compass displays one stone for each direction, however, others of like color can be utilized. Using one of the others does not imply a lesser energy or connection with that particular direction, season, or element. A selection of gemstones is essential in order to employ the one that works best for you. Everyone's energy patterns are unique and react differently not only to different types of gemstones, but also to individual stones. This is why it is important to explore a variety of gemstones and crystals when purchasing them.

Use this compass in the same way as the pa tzu compass to identify the gemstone, direction, and energy group that holds the most influence for you. Find your lo shu number on the gemstone compass (see Table 4.2, page 44). For example, if your compass

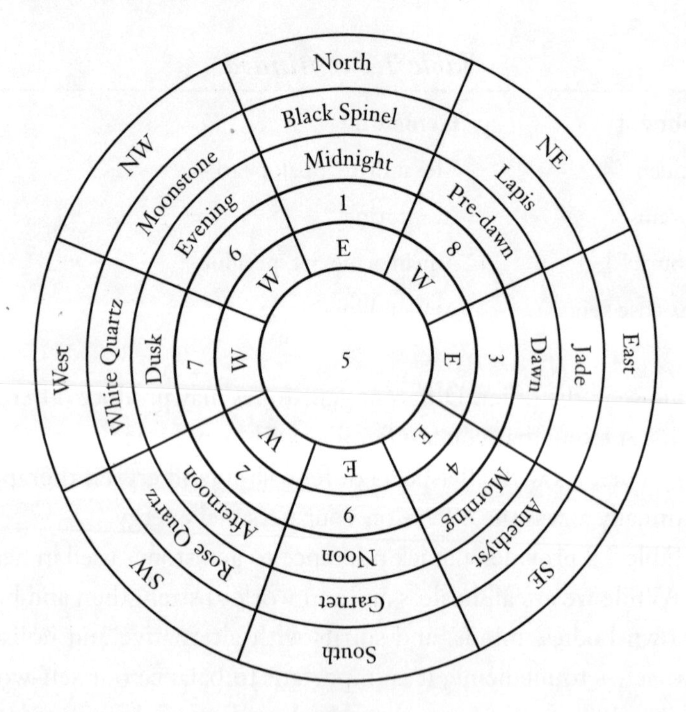

Figure 7.3
Gemstone pa tzu compass.

number is one, your power direction is North and the gemstone black spinel holds a strong influence for you. You are likely to find that you feel most comfortable in a house seated in (has its back toward) the North and you may choose to wear black frequently. Take time to explore how you feel about your direction, stone, and other aspects associated with it.

As was discussed in chapter 4, also associated with your compass direction are your positive and negative directions (see Figures 4.8 and 4.9, pages 53–54). Gemstones can be employed to help you deal with issues that may arise from the location of these directions in your home. For example, in Figure 4.10 (page 56) the Misfortune direction (Northwest) was located in the family's kitchen, which could foster illness and a lack of vitality. The occupants

Table 7.3—Gemstones to Aid in Handling Negative Directions

Gemstone	Associated Powers
Agate	General healer; good for grounding and balance; aids in attracting abundance and luck; promotes longevity and strength; provides protection
Alexandrite	Helps achieve success; fosters good luck
Amazonite	Promotes communication, clarity, and trust; disperses negative energy
Amber	General healer; good for balance and calming; associated with wisdom and knowledge
Amethyst	General healer—physical and spiritual; calms and transforms
Andalusite	Attracts success; builds leadership
Apatite	Promotes concentration; aids the intellect; promotes harmony
Aquamarine	Produces mental clarity; aids in dealing with loss, grief, and fear
Aventurine	General healer; enhances success in career; aids creativity
Azurite	For cleansing and spiritual guidance; promotes patience
Beryl	Stimulates communication and acceptance
Bloodstone	Attracts good luck and abundance
Carnelian	Aids creativity and self-worth
Chalcedony	Alleviates melancholy; builds vitality
Chrysoberyl	Fosters benevolence and optimism
Chrysoprase	Aids communication and stability
Citrine	Boosts creativity and self-worth; attracts prosperity and stability; aids protection

Table 7.3 continued

Gemstone	Associated Powers
Coral	Clears negativity; stimulates relationships; provides protection
Diamond	Builds relationships; attracts abundance; supports longevity
Emerald	Provides insight; a healer that helps navigate difficulties
Fluorite	Provides protection and strength in times of transition
Garnet	Promotes confidence and success
Iolite	Builds self-confidence and inner strength
Jade	General healer; promotes longevity and wisdom
Jet	Engenders honor and justice; provides protection during transitions
Kunzite	Provides emotional support
Lapis lazuli	Promotes inner power and tranquility; expands understanding
Malachite	Attracts loyalty and comfort; aids in navigating setbacks and difficulties
Moonstone	Alleviates fear; balances yin and yang; promotes protection through insight
Obsidian	Provides grounding and insight; dispels half-truths
Onyx	Helps control emotions and negative thoughts
Opal	Stimulates a wider vision
Pearl	Induces emotional balance and openness
Peridot	General healing—especially hurt feelings; attracts comfort; builds vitality

Table 7.3 continued

Gemstone	Associated Powers
Quartz	General healing and balancing—emotional and physical; alleviates anger; reveals distortion
Rhodochrosite	Provides support during transitions; attracts comfort
Rhodonite	Provides emotional support through wider vision
Ruby	Strengthens self-esteem and integrity; attracts and engenders generosity; dispels fear
Sapphire	Promotes mental clarity and intuition
Spinel	General healer; enhances ability to overcome obstacles and setbacks; enhances communication
Tanzanite	Aids in dealing with changes and weathering difficulties
Topaz	Alleviates tension; promotes communication; attracts abundance
Tourmaline	Aids in dealing with grief; dispels fear of positive change; provides protection from negative energy
Turquoise	General healing; protects against negativity
Zircon	Attracts prosperity and abundance

would not want to employ the use of moonstone or any gray stone there, because it would strengthen the negative power of this direction. Instead, use a gemstone such as bloodstone, which is a stone of strength, to neutralize toxins and attract good luck. Table 7.3 provides a quick reference for employing gemstones to enhance your positive directions and combat the potential problems of negative directions.

Analyze your negative directions to determine whether or not they may be affecting areas of your life. These directions only indicate potentials, and it is important to refrain from looking for problems that may not exist. However, if there are issues in your life that need attention, take time to examine them. If a setback or problem seems to be caused by a lack of communication or miscommunication, introduce a piece of chrysoprase or spinel into the Difficulty or Setback direction of your room or home. In the Loss or Misfortune areas, you may want to employ a gemstone that promotes protection. If these areas contain the desk or table where you handle your finances and there is no other place to locate it, try using a gem that attracts prosperity.

CHAPTER EIGHT

Gemstones and Birthdays

Birthstones

People who have not heard of feng shui or crystal therapy can usually tell you their birthstone. The common American jeweler's list, compiled in 1912, popularized the use of certain gemstones in mainstream culture. Before this list was adopted, varying forms of older lists were used. In 1937 British jewelers adopted their own list to "correct" duplicates in the older ones. Both of these lists became standard in their respective countries. The primary intention of creating the lists was to get rid of conflicts in older lists of birthstones. Also, the list of zodiac gemstones straddled the months and different sources could not agree on when a zodiac sign began and ended. Today the jeweler's lists have been expanded to include alternates, and other lists have been developed to accommodate fashion trends. While we may seem to be back where we were a century ago, these various lists offer flexibility. Birthstones may have an influence on us, however, as more people become attuned to their own energy, they find that one size does not fit all. Having said that, the variety of birthstones in all their forms offer a place to start for exploration.

The modern jeweler's lists evolved from the idea that gemstones are more powerful at certain times of the year and endow the person who wears them with luck and good fortune. Ancient

Table 8.1—Western Color Associations

White	Purity, innocence, truth
Blue	Heaven (spiritual), devotion, virtue
Red	Love, passion, power
Green	Growth, hope, faith
Purple	Sadness, suffering
Yellow	God's love, faith

people were observers of the natural world. They were in awe of it and appreciated its mysteries. Through these observations, colors became symbolic of life. The ancient Chinese incorporated this into their system of feng shui. In Western cultures this adaptation was slightly less formal. The symbolism of these color assignments were more readily utilized by the Catholic Church. Think of the robes of bishops (red) and the predominant color used to celebrate Easter (purple) (Table 8.1).

Color symbology and use of gemstones by religious leaders is documented in the Bible on the religious caftan or breastplate of Aaron, brother of Moses (Exodus 28:15–30). The twelve stones that adorned this robe represented the twelve tribes of Israel. The Roman historian Flavius Josephus translated the list of gemstones into what was known and popular in his time. In 1913, George F. Kunz, gemologist to Tiffany & Company, offered a "correction" of Josephus's list.[1] Another set of twelve gemstones is mentioned in Revelation 21:19–21 as the foundation stones for the wall of heavenly Jerusalem. See Table 8.2.

Most of the stones in both of the biblical lists are the same, especially if one takes into account that gemstones were given different names and were frequently mistakenly identified. For example, *topaz* was a name applied to a wide range of yellowish

Table 8.2—Comparison of Biblical Gemstones

Gemstones in the breastplate of Aaron	**Flavius Josephus's translation**	**George Kunz's "correction"**	**Foundation stones for heavenly Jerusalem**
Sardius (carnelian)	Sardonyx	Carnelian	Jasper
Topaz	Topaz	Chrysolite (peridot)	Sapphire
Carbuncle	Garnet	Emerald	Chalcedony
Emerald	Emerald	Ruby	Emerald
Sapphire	Sapphire	Lapis lazuli	Sardonyx
Diamond	Diamond	Onyx	Sardius (carnelian)
Ligure	Amber	Sapphire	Chrysolite (peridot)
Agate	Agate	Agate	Beryl
Amethyst	Amethyst	Amethyst	Topaz
Beryl	Aquamarine	Topaz	Chrysoprasus
Onyx	Onyx	Beryl	Jacinth (zircon)
Jasper	Jasper	Jasper	Amethyst

stones, as well as to olivine, which is green. Likewise, *carbuncle* referred to red stones.

At some point in the distant past the twelve gemstones mentioned in the Bible were ascribed to the months. Since the gemstones already represented the tribes of Israel and color symbology was in use, it would not take a great leap of the imagination to associate the stones with the months of the year—both conveniently number twelve. If the idea of a special gemstone talisman based on your birth appeals to you, but you have not been attracted to stones listed for you by jewelers, you may find one of ancient origin that suits you better. When I was growing up, I'd only heard that topaz was the stone of my birth month, but it didn't do anything for me. I was attracted to citrine and later discovered that this is also considered a November stone. Lists evolve and different lists may not agree; the "modern" list included in Table 8.3 was compiled from various sources.

Over the centuries, other lists were created that were national in origin. One of these was created by Isidore, the bishop of Seville, in the year 635.[2] Other stones were generally used throughout Western Europe from the fifteenth to twentieth centuries. Explore what is prescribed for your birth month in Table 8.4, but ultimately go with what your heart tells you. Note that as in other lists, there is an overlap in the names applied to some gemstones. Chrysolite (peridot) is a type of olivine. Aquamarine is a type of beryl.

Gemstones and Astrology

Early astrologers believed that stars associated with the twelve figures of the zodiac (the signs that mark the sun's and moon's paths across the sky) emitted special energy, which affected life on Earth. They also believed that certain gemstones attracted the pure vibrations of these heavenly bodies. The gemstone corresponding to a person's zodiac sign acted like a lightning rod for this cosmic energy. Also, the use of talismans was popular with the ancient Romans who believed that the healing power of a stone could be amplified when engraved with an astrological sign. This, too, enhanced a gemstone's ability to filter out negative energy and protect the wearer.

Vedic astrologers of India assigned gemstones to the seven planets in their methodology (the outer planets of our solar system were not known at the time). The planetary stones have color

Table 8.3—Birthstones of Modern and Ancient Traditions

Month	Modern	Arab	Hebrew	Hindu	Roman
January	Garnet, rhodolite, rose quartz, rubellite	Garnet	Garnet	Ruby	Garnet
February	Amethyst, onyx, moonstone	Amethyst	Amethyst	Topaz	Amethyst
March	Aquamarine, blue topaz, bloodstone	Bloodstone	Bloodstone, jasper	Opal	Bloodstone
April	Diamond, zircon, beryl, white sapphire	Sapphire	Sapphire	Diamond	Sapphire
May	Emerald, tourmaline, tsavorite, garnet	Agate	Agate, carnelian	Emerald	Agate
June	Pearl, moonstone, alexandrite, opal, cat's-eye chrysoberyl	Agate	Emerald	Pearl	Emerald

Table 8.3 continued

Month	Modern	Arab	Hebrew	Hindu	Roman
July	Ruby, spinel rubellite,	Carnelian	Onyx	Sapphire	Onyx
August	Peridot, sardonyx, tourmaline, emerald, jade	Sardonyx	Carnelian	Ruby	Carnelian
September	Sapphire,blue spinel, iolite, lapis, blue tourmaline	Peridot	Peridot	Zircon	Peridot
October	Opal, garnet, tourmaline, sapphire, kunzite, morganite	Aquamarine	Aquamarine, tourmaline	Coral	Aquamarine
November	Topaz, citrine, beryl, chrysoberyl, yellow sapphire	Topaz	Topaz	Cat's-eye chrysoberyl	Topaz
December	Turquoise, blue topaz, zircon, aquamarine	Ruby	Ruby	Topaz	Ruby

Table 8.4—Birthstones by Nation

Month	Italy	Poland	Russia	Seville	General
January	Hyacinth, garnet	Garnet	Garnet, hyacinth	Hyacinth	Garnet
February	Amethyst	Amethyst	Amethyst	Amethyst	Amethyst, hyacinth, pearl
March	Jasper	Bloodstone	Jasper	Jasper	Jasper, bloodstone
April	Sapphire	Diamond	Sapphire	Sapphire	Diamond, sapphire
May	Agate	Emerald	Emerald	Agate	Emerald, agate
June	Emerald	Agate	Agate	Emerald	Agate, cat's-eye chrysoberyl, turquoise

Table 8.4 continued

Month	Italy	Poland	Russia	Seville	General
July	Onyx	Ruby	Ruby	Onyx	Turquoise, onyx
August	Carnelian	Sardonyx	Alexandrite	Carnelian	Sardonyx, carnelian, moonstone, topaz
September	Olivine	Peridot	Olivine	Olivine	Chrysolite
October	Beryl	Aquamarine	Beryl	Aquamarine	Opal
November	Topaz	Topaz	Topaz	Beryl, topaz	Topaz, pearl
December	Turquoise, ruby	Turquoise	Turquoise, chrysoprase	Ruby	Ruby, bloodstone

Table 8.5—Gemstone Associations with the Planets and Stars

Zodiac Sign	Stone	Lucky	Planetary Planet	Stone
Capricorn Dec. 22–Jan. 20	Ruby, agate, garnet, turquoise, smoky quartz, beryl	Ruby, onyx	Saturn	Lapis lazuli, sapphire
Aquarius Jan. 21–Feb. 21	Garnet, moss agate, opal, amethyst	Jasper	Saturn	Turquoise, sapphire
Pisces Feb. 22–Mar. 21	Amethyst, sapphire, bloodstone, jade, aquamarine, diamond	Ruby	Jupiter	Aquamarine, yellow sapphire
Aries Mar. 22–Apr. 20	Bloodstone, diamond, ruby	Topaz	Mars	Jasper, red coral
Taurus Apr. 21–May 21	Sapphire, turquoise, amber, emerald, coral	Garnet	Venus	Emerald, aventurine

Table 8.5—continued

Zodiac Sign	Stone	Lucky	Planetary Planet	Stone
Gemini May 22–Jun. 21	Agate, pearl, chrysoprase, aquamarine	Emerald	Mercury	Emerald
Cancer Jun. 22–Jul. 22	Emerald, agate, moonstone, pearl, ruby	Sapphire	Moon	Pearl, chalcedony
Leo Jul. 23–Aug. 22	Onyx, tourmaline	Diamond	Sun	Amber, ruby
Virgo Aug. 23–Sep. 22	Carnelian, jasper, jade, sapphire	Turquoise, zircon	Mercury	Emerald
Libra Sep. 23–Oct. 22	Peridot, opal, lapis lazuli	Agate, beryl	Venus	Chrysolite, diamond
Scorpio Oct. 23–Nov. 21	Topaz, beryl, coral	Amethyst	Mars	Aquamarine
Sagittarius Nov. 22–Dec. 21	Topaz, amethyst, turquoise	Beryl, pearl	Jupiter	Yellow sapphire

Table 8.6—Gemstone Associations with the Days of the Week

Day	Gemstone
Monday	Moonstone, pearl
Tuesday	Emerald, ruby, sapphire
Wednesday	Amethyst, lodestone, ruby
Thursday	Carnelian, cat's-eye chrysoberyl, sapphire
Friday	Alexandrite, cat's-eye chrysoberyl, emerald
Saturday	Diamond, labradorite, turquoise
Sunday	Sunstone, topaz

Table 8.7—Gemstone Associations with the Hours of the Day

A.M. Hours	Gemstone	P.M. Hours	Gemstone
1	Quartz	1	Zircon
2	Hematite	2	Emerald
3	Malachite	3	Beryl
4	Lapis lazuli	4	Topaz
5	Turquoise	5	Ruby
6	Tourmaline	6	Opal
7	Chrysolite	7	Sardonyx
8	Amethyst	8	Chalcedony
9	Kunzite	9	Jade
10	Sapphire	10	Jasper
11	Garnet	11	Lodestone
12	Diamond	12	Onyx

wave lengths similar to their prescribed planets. Throughout the ages, people have been fascinated by the heavens and the influence other planets (and stars) have on us. Associating gemstones with heavenly bodies allows a person to remain grounded while his or her spiritual life soars.

Because traditions vary, there are multiple stones associated with each zodiac sign. In addition to a stone that taps into the astrological powers, there are also stones that are considered lucky to people born under each sign. See Table 8.5.

Ancient astrologers also assigned gems as talismans for certain days of the week, as well as hours of the day. See Tables 8.6 and 8.7.

In both the Hebraic and Christian traditions, guardian angels got into the act and were assigned to the zodiac as well as specific months and gemstones. (One wonders if people were doubly protected by multiple angels where months and zodiac signs overlapped.) See Table 8.8.

Chinese Astrology

Extensive study of the Lo Shu Grid (Magic Square) found on the back of the tortoise (discussed in chapter 4) gave rise to not only a school of feng shui and the *I Ching*, but also Chinese astrology. The full lou pan (compass) of the Compass School of feng shui includes aspects of astrology that are beyond the scope of this book. However, it is important to note that astrology is integrated with feng shui, and the use of birthstones has a place in personal feng shui work. In Chinese astrology, particular astrological signs (which also number twelve) are associated with the cardinal directions. Unlike the Western zodiac, which cycles through all twelve signs in one year, it takes twelve years to cycle through all of the Chinese signs. Table 8.9 includes the angels, and Chinese and Western signs associated with the cardinal directions.

If all these variations seem confusing, take some time to explore gemstones that might have an influence in your life. If you

Table 8.8—Guardian Angel Associations

Angel	Gemstone	Zodiac	Month
Adnachiel	Beryl	Sagittarius	December
Ambriel	Garnet	Gemini	May
Asmodel	Topaz	Taurus	April
Barbiel	Amethyst	Scorpio	November
Barchiel	Jasper	Pisces	February
Gabriel	Onyx	Aquarius	January
Hamaliel	Zircon	Virgo	September
Machidiel	Ruby	Aries	March
Muriel	Emerald	Cancer	June
Uriel	Agate	Libra	October
Verchiel	Sapphire	Leo	July

Table 8.9—Astrological and Angelic Associations of Cardinal Directions

Direction	Western	Chinese	Angel
East	Aries	Dragon	Mikhael
South	Capricorn	Phoenix	Uriel
West	Libra	Tiger	Raphael
North	Cancer	Tortoise	Gabriel

feel connected to a particular birthstone (by zodiac or month), use it to ground and center yourself for feng shui work. If an hour stone or day of the week stone (determined by the hour and day of your birth) works better for you, use that instead. You may find that a combination of stones supports your personal energy. If so, use them together. Birthstones, in all their forms, are symbols of the spirit of cycles.

Birthstone Meditation

While it is not necessary for gemstone feng shui work, an elemental/seasonal meditation with your birthstone(s) can be a learning experience. This meditation will help you tune into the effects that the five feng shui elements, the directions, and seasons have on you. Your birthstone(s) will hold a comfortable vibration, and the frequency of the elemental and seasonal stones will interact with your birthstone(s). The meditation is similar to the elemental/seasonal meditation described in chapter 2, but the difference is the use of the five feng shui elements (instead of the four Western elements), and, of course, the stones that represent the seasons and directions.

With a magnetic compass, determine which direction is North in order to place the stones in the proper locations. Lay the seasonal and directional stones—black spinel, lapis lazuli, jade, amethyst, garnet, rose quartz, white quartz, and moonstone—to create the outer ring on the floor. Make the circle large enough to sit or stand in the middle. Lay the elemental stones—peridot, opal, malachite, jet, and andalusite—beside the others toward the center of the circle. Place a citrine and third piece of andalusite in the middle under your chair (or be careful not to step on them if you choose to stand). See Figure 8.1.

Holding your birthstone(s), begin by facing the stones nearest to your birth month. Take a couple of deep breaths, relax, and allow yourself to feel the energy that the three sets of stones—direction, element, and season—produce. Take as much time as you feel appropriate with each direction. When you have come full circle, take time to reflect on Center and earth. When you feel grounded, open your eyes and take a moment to review what you may have felt.

Becoming familiar with the elements and directions in this way will help you more easily identify energy issues in your feng shui practice. You may also find this meditation calming and useful for

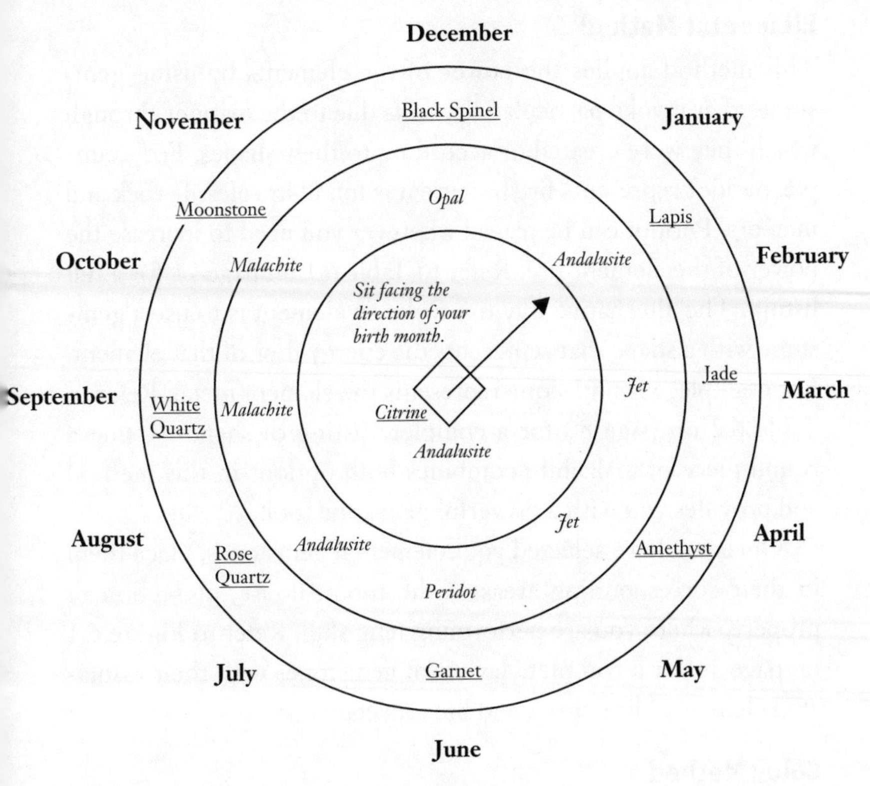

Figure 8.1
Elemental/seasonal meditation with birthstones.

getting in touch with your own energy even when you are not preparing for feng shui work.

In Conclusion

Experiment with the different methods for employing gemstones in your feng shui practice. You may want to try each individually and then in combination to see what works best for you. Any of these methods can be the basis for a gemstone bagua that can be used to activate healthy energy wherever it is needed.

Elemental Method

This method applies the power of the elements by using gemstones that invoke particular elements due to the manner through which they were created or according to their shapes. For example, peridot represents fire because it is found in volcanic rock and meteors. Peridot can be placed wherever you need to increase the power of the element fire. Refer to Table 6.1 on page 71 for a full listing. The alternative way to invoke an element is to use a gemstone with a shape that represents the energy flow of that element. For example, a round stone represents the element metal. Refer to Table 6.2 on page 82 for a complete listing of shapes. Using a round piece of malachite combines both options in this method and provides you with a powerful gemstone tool.

Once you have selected your elemental gemstones, place them in their corresponding areas of the room, house, or section of property where you are performing feng shui. Refer to Figure 6.1 on page 71 for a complete layout of gemstones with their associated elements, directions, and life aspects.

Color Method

This method raises the power of a direction by employing a stone with the color associated with that direction. For example, Northwest is represented by the color gray, so using moonstone or any other gray gemstone would be appropriate. Refer to Figure 7.1 on page 83 for a complete layout of gemstones with their associated directions and life aspects. Parts of the body are also associated with sectors of the Magic Square and can benefit from the use of feng shui. Refer to Figure 7.2 on page 86 for a complete layout of gemstones associated with parts of the body. Working on these related areas in your home can be done in conjunction with crystal therapy on your body. Refer to Table 7.2 on page 87 for a listing of common ailments and their gemstone remedies.

Gemstones can also be used to counteract potential issues related to your negative directions. For example, if your lo shu number is four, Difficulty occurs in the West sector for you, which is the area associated with creativity. Refer to Figures 4.8 and 4.9 on pages 53–54 to find your negative and positive directions. You may try placing citrine in the West sector of the room (or wherever you are performing feng shui) because this gemstone boosts creativity and can help counteract negative energy that could inhibit your creative processes. You could also use amazonite since this stone aids in dispersing negative energy. Refer to Table 7.3 on page 91 for a listing of gemstones and their associated powers.

Birthstone Method

Birthstones can add a personal dimension when creating or attracting healthy energy if your birth month corresponds with the same direction as an element. For example, if you were born in December, this month corresponds with North. Black spinel and opal gemstones can be used in this area to enhance your connection with North and the element water. If this is your power direction or one of your power points, this would be an exceptionally potent direction for you. In contrast, if your birth month corresponds with one of your negative directions, your birthstone can be influential in combating the potential issues and challenges associated with these directions. Refer to Figure 8.1 on page 109 for an overview of how gemstones relate to the months of the year.

---

1. Kunz, *The Curious Lore of Precious Stones*, 319.
2. Baur and Boušk, *A Guide in Color*, 228.

PART TWO

A Compendium of Gemstones

This part of the book provides an overview of the most common gemstones and minerals, including their history, associations, and use in feng shui. There is a noticeable similarity among many of the names' origins, especially among Latin, Greek, and Sanskrit (the language of Hindu law and religion). This is because these three evolved from the common mother tongue Proto-Indo-European.

While I have included a range of information here, this is by no means a compilation of complete material; that would take its own volume and then some. If you sense that a stone is providing or supporting energy in a way not described here, trust what you feel and check other sources for further information.

Agate

Agate is a type of chalcedony quartz that took its name from the River Aghates near Sicily; it is now called River Drillo. As an aid to emotional healing, agate is said to help discern the truth as well as accept circumstances. Because agate was believed to help protect against high fevers, it was placed in drinking water to banish illness. Agate was also thought to be instrumental in relieving arthritis, headaches, and thirst.

In addition to being worn as an amulet, agate was used for decorative jewelry and small objects. During the classical era in Greece and Rome, agate was widely used for cameos and intaglios. These were frequently worn upside down for the wearer's enjoyment.[1] Bowls carved of agate were common throughout the Byzantine Empire, and Mithridates had amassed a collection of several thousand.[2] During the Renaissance in Europe, collecting agate bowls surged in popularity again as did its use in jewelry and furniture inlay.

Not only does agate come in a wide variety of colors, but it also has a plethora of fascinating effects. Eye agate actually looks like an eye, and plume agate displays a feathery pattern. Because of layers that form when agate is created, some colors and textures are structured into bands of color while others seem to create "scenes"—call it nature's Rorschach test. There are also the delicate-patterned lace agates.

Agate is good for grounding and balance. It fosters abundance, wealth, love, and connection with the natural world. It is also used for protection. Moss agate attracts abundance and aids in building self-confidence and strength. Lace agate helps to clear the mind.

Color(s) A wide range and usually multicolored; black-banded, blue-lace, crazy-lace, moss, tree, and white

Associations General—Gemini; solar plexus chakra; Earth, Mercury; yin/yang
Moss agate—Virgo; heart chakra; yin
Lace agate—Pisces; throat and third-eye chakras; yin

Feng shui use Center (balance/connection to the natural world); Northeast (wealth); Southwest (love); negative direction Misfortune (protection/combat illness)

Alexandrite

Discovered in Russia in 1830 and named for Czar Alexander II, this gemstone of the chrysoberyl family changes color according to the light—natural or incandescent—in which it is viewed. Because of the green/red color change, alexandrite was fashionable with the aristocracy in Russia where the imperial colors were red and green. In the rest of Europe, alexandrite was especially popular in the late nineteenth and early twentieth centuries.

While alexandrite has been somewhat rare and expensive, another source was discovered in Brazil in the late 1980s, which gave this gemstone a newfound popularity. Color-changing sapphire grown in a lab and sometimes called *alexandrine* is used as an imitation.

As a healing stone, the use of alexandrite with other stones enhances the power of the others. On its own, alexandrite strengthens personal power and spirituality. It is also effective in building intellect and creativity. It is a stone for achieving success and has been used as a good luck amulet. (*See also* Chrysoberyl.)

Color(s) A color-changing stone—greens/blues to reds/violets

Associations Scorpio; crown chakra

Feng shui use Center (spirituality); South (success and reputation); Southeast (self-/net worth); any negative direction (good luck amulet); any area where healing is needed

Amazonite (Microcline)

Although this type of feldspar takes its popular name from the Amazon Basin where it was first thought to be jade, amazonite has been widely used for thousands of years. This gemstone's older name, *microcline*, comes from the Greek words *micro* for "small" and *klino* meaning "tilt." It is suspected that this second word was associated with it because of the shape of its crystals.

Amazonite was used for jewelry in Egypt, Mesopotamia, India, and the Sudan as early as 2000 B.C.E. It was included among the gemstones found in King Tutankhamen's tomb. While the Egyptians used this gemstone widely for amulets, tablets of amazonite have been found with part of the *Egyptian Book of the Dead* engraved upon them. In pre-Columbian Central and South America, amazonite was used for personal adornment. Ancient Assyrians believed amazonite was the gemstone of their god, Belus, and used it in sacred rituals.

Amazonite is useful to disperse negative energy. It is also emotionally soothing and can aid in communication with a lover. It inspires openness, trust, and honor. (*See also* Feldspar.)

Color(s) Green, blue-green

Associations Virgo; heart and throat chakras; yin

Feng shui use Northeast (self-cultivation); Southwest (partner/relationships); any negative direction (disperse negativity)

Amber

Now sometimes referred to as the "Jurassic gem," the film *Jurassic Park* has made amber popular again. The age of a piece of this organic gem can be anywhere between 1 and 360 million years. Amber was created by heat and pressure applied to sticky tree (usually pine) resin. The presence of insects and leaves in amber adds to its value on the market.

Amber's long history includes a lengthy association with humans. Archaeologists have found artifacts made of amber that date to 8000 B.C.E. It was also used by the ancient Assyrian, Egyptian, Phoenician, and Greek civilizations. Labeled the "gold of the north" by some, the Greeks called it *electron*. Rubbing a piece of amber with cloth will build up an electric charge that makes it attract small pieces of paper, feathers, and dust.

Because of amber's most common color, it has been linked with the sun. In Ovid's *Metamorphoses*, the ancient tale of Phaeton (the son of Phoebus, the sun) tells of his death and how his mother's tears dry into pieces of amber.[3] This gemstone is also mentioned in Homer's *Odyssey* and the writings of Pliny. Only upper-class Romans could afford amber, however, gladiators lucky enough to obtain it wore it as an amulet for its protective powers.

Because it gives off a pleasing odor when burned, amber was utilized in temples throughout Asia. In medieval Europe, amber was used for rosary beads. During this period it was also worn to ward off disease. During the eighteenth century, amber was used to adorn rooms; it was inlaid in wall panels as well as door and window frames.

Amber is useful for yin/yang balance because it draws on the power of the sun while keeping the wearer grounded. It is also a healing and protective "gem" that attracts luck, calms energy, and aids in building vitality. Amber is also useful to manifest general change, and inspire love and wisdom.

Color(s) Colorless/white, pale yellow to dark brown, reddish, green, black, blue

Associations Leo and Sagittarius; sacral chakra; Mercury, Sun; yin/yang

Feng shui use Secondary gemstone for the element wood; Center (balance/calm); Northeast (wisdom); West (ancestors); Southwest (love); negative directions Misfortune and Loss if these fall in your kitchen or dining room (builds vitality); any direction to help boost changes you want to attract into your life

Amethyst

A variety of quartz, this gemstone has enjoyed a long history of popularity that has rarely waned. Since ancient Egypt it has been a prized stone of royalty. It was preferred by Catherine the Great of Russia and has an honored place in the crown jewels of England. Amethyst was especially popular during the Renaissance. Pale-colored amethyst was occasionally called "Rose de France" and was frequently used in jewelry during the Victorian era.

Amethyst was a favored stone worn by bishops in the Middle Ages. Since the sixteenth century, each new pope is given an amethyst ring during his investiture. This ring is destroyed when the pontiff dies and a new one is made for his successor.

This gem's name comes from the Greek word *amethustos*, as well as the Latin *amethystus*, which translates as "not drunken" or "without wine." Throughout the ages amethyst has been hailed as a preventative for getting drunk. Goblets carved from amethyst were believed to allow drinkers to imbibe as much as they liked without suffering the aftereffects. It is suspected that the drinker could appear to be enjoying vast quantities of wine while actually consuming water that looked wine-colored through the amethyst. This could also be the source of the belief that amethyst aids in overcoming alcoholism.

Several versions of a Greek myth tell of a young woman named Amethyst who, while on her way to pay tribute to the goddess Diana, is set upon by tigers dispatched by an angry Dionysus (or Bacchus). Diana turned Amethyst into a statue of white quartz to protect her from the tigers' claws. One version of the story tells that the remorseful tears of Dionysus (which were wine, of course) fell on the statue and turned it into a purple quartz. The other version merely states that a less-charitable Bacchus poured wine over the statue. This story could be the root of the belief in amethyst's powers of protection.

Today amethyst is commonly used in crystal work to transform body energy into the spiritual realm. As a gemstone of change, amethyst is a general healer, spiritual tool, and a stone of wisdom. It is good for calming the mind and attracts good luck as well as love.

Color(s)	Pale lilac to deep purple
Associations	Pisces; third-eye chakra; Jupiter, Pluto, Neptune; yin/yang
Feng shui use	Primary gemstone for the Southeast direction; Center (balance/spiritual growth); Northeast (wisdom); negative directions (protection/good luck); anywhere to help manifest change

Ametrine

Found only in Bolivia, ametrine was presented on the world stage in the seventeenth century by a Spanish conquistador who brought gifts from the New World to his monarch. This dual-colored quartz combines the deep beauty of amethyst with the light sunshine of citrine. Although the colors of amethyst and citrine are opposite to each other on the color wheel, this pairing of complementary opposites provides yin/yang balance.

Ametrine brings the physical and spiritual realms into balance. It calms negative emotions and is used to cleanse the aura.

Color(s) Dual-colored lilac/purple and yellow

Associations Libra; all chakras; yin/yang

Feng shui use Center (balance/spiritual growth); any negative direction (cleansing)

Andalusite

Andalusite is called the "earth stone" because of its soothing earth-tone colors and grounding vibrations. It gets its name from Andalusia, the area in Spain where it was discovered. Unlike other multicolored gemstones that display bands or speckling, andalusite is a pleochroic gem that presents its colors through a dance of patterns. Usually a cutter will try to bring out the best color in a stone and minimize the others, however, andalusite's strong pleochroism is encouraged to produce an exciting play of colors. At one time andalusite was known as "poor man's alexandrite," which was a misnomer. Alexandrite's color change depends on the light in which it is seen, whereas andalusite's depends on the angle of view.

A variety of andalusite called *chiastolite* often displays a dark cross pattern when sliced. These were found frequently near Santiago de Compostela in Spain, which is an important Christian pilgrimage site. Chiastolite became known as the "cross stone" and was sold to pilgrims.

Kyanite is formed from andalusite. An increase in pressure or decrease in temperature will cause andalusite to recrystallize into this mineral. Andalusite has seen industrial use as a mineral ingredient in the manufacturing of spark plugs and special porcelains.

Andalusite's "earth stone" reputation is upheld in its use for grounding and balance. Like Mother Earth, andalusite touches the emotions and aids in the spiritual journey. This gemstone fosters wise leadership and success.

Color(s) Brown-reds, greens, yellows

Associations Virgo; root chakra; Earth; yin

Feng shui use Primary gemstone for the element earth; Center (grounding/balance/spiritual growth); North (success); anywhere to help manifest change

Apache Tears

See Obsidian's entry for a full description. Apache tears are used for protection and to attract good luck.

Color(s) Black

Associations Saturn; yang

Feng shui use Any negative direction (protection/luck)

Aquamarine

This variety of beryl gets its name from the Latin *aqua marina*, meaning "sea water." The stone's blue-green color is reminiscent of the soothing water of the Mediterranean. Mythology says that this gem was presented as a gift by King Neptune to the mermaids. Sailors have used it as a protective amulet as it was believed to possess its greatest degree of strength when in water. It also symbolized the moon, a connection which is easy to understand since the moon affects the Earth's tides. Its earliest documented use was in ancient Greece.

Natural aquamarine is more green than blue, depending on the amount of iron in it. It is usually heated to subdue the green. Light blue-green synthetic spinel is used to imitate aquamarine.

This gemstone is said to bring love to those who wear it. Its comforting influence on couples promotes fidelity and calms differences. Aquamarine's power to moderate supports and enhances good communication. It is also known as the "stone of courage." Since it imparts courage to the wearer, it also offers protection. Aquamarine promotes cleansing and tranquility, especially when dealing with loss or grief. (*See also* Beryl.)

Color(s) Light-blue, green-blue, blue

Associations Gemini, Pisces, Aries; throat and heart chakras; Neptune, Moon; yin

Feng shui use Southwest (relationships/love); negative directions (courage/protection)

Aventurine

This gemstone is a type of sunstone, which is a variety of feldspar. Tiny flecks of mica or hematite produces its iridescent sheen. In China it was called "Imperial Yü." Aventurine began its current wave of popularity in the nineteenth century.

This gemstone is a powerful healer both physically and emotionally, and has been called the "healer of heart and soul." It quells anxiety, aids in finding solutions to life's problems, and assists in making the right choices. Aventurine attracts career success and fosters creativity. It is also a calming gemstone.

Color(s)	Mostly green, also red-brown, occasionally blue
Associations	Aries; heart chakra; Mercury, Venus, Uranus; yang
Feng shui use	North (career); West (creativity); any area/life aspect that needs an emotional lift; any negative direction (solutions/choices/luck)

Azurite

This gemstone's 55 percent copper content is responsible for its deep color. Azurite frequently forms with malachite; both are created by the oxidation of a copper such as chalcopyrite.

For centuries azurite has been used as a pigment for fabric and paint. Because of its color, azurite has been dubbed the "stone of heaven." Ancient Mayans revered azurite for its ability to help the wearer connect with wisdom. In Native American traditions it is believed to help the wearer contact his or her spirit guides.

Azurite aids in strengthening intuition and communication. It also helps to unseat and release deep-rooted problems. Azurite promotes patience, cleansing, and spiritual guidance.

Color(s)	Azure, pale blue to deep blue
Associations	Sagittarius; throat and third-eye chakras; Saturn, Venus; yin
Feng shui use	Center (spiritual growth/guidance); Northeast (wisdom); negative directions (communication/remove problems)

Benitoite

As the official gemstone of California, the only significant source of benitoite is in the San Benito county of that state. In Spanish, *benito* means "blessed." This gemstone was first discovered in 1907, however, there is controversy as to who was the first to unearth it: the team of Hawkins and Sanders (who at first thought they'd found sapphire), or James Marshall Couch. Benitoite was the first example of the ditrigonal-dipyramidal crystal shape to be found. It is easily confused with sapphire. This gemstone promotes understanding on an emotional level.

Color(s)	Blue, blue/violet, black (colorless and pink are extremely rare)
Associations	Libra; heart and third-eye chakras; Venus; yang
Feng shui use	Use as supportive gemstones in the Magic Square sectors/directions for their respective colors.

Beryl

Beryl comes from the Greek word *beryllos*, which was used to identify the gemstones known as beryl as well as most other types of green stones. Beryl is a group of minerals that includes aquamarine, emerald, heliodor, morganite, and others. The many colors of beryls are the result of varying amounts of metals within these minerals.

Various beryls were used as cutting tools during the Upper Paleolithic period, however, the first technical use, recorded by Pliny, is the cut emerald Emperor Nero used as a monocle. Because of the size of a stone required to provide a slice for such use, later scholars believed that the gem was actually an aquamarine.[4]

Morganite was named for the nineteenth-century industrial baron J. P. Morgan who was an avid gem collector.[5] This name was applied by Tiffany's George Kunz to honor Tiffany & Company's best customer when it was discovered in California. Morganite, however, was known and mined elsewhere in the world and was popular from the seventeenth through nineteenth centuries.

Other types of beryl include heliodor and bixbite. Heliodor has been called the "gift of the sun" as its name implies from the Greek words *helios* ("sun") and *doron* ("gift"). In medieval Europe, heliodor was believed to "cure" laziness. Bixbite is named for Maynard Bixby who cataloged the minerals of Utah, however, this is not a scientifically recognized type of beryl.

Beryl was one of the gemstones in the breastplate of Aaron in the Bible. This gemstone stimulates communication, acceptance, and healing. It also supports spiritual growth.

Color(s) Gold, yellow, green, pink (colorless and red are relatively rare)
Bixbite—strawberry/raspberry color
Golden beryl—lemon to golden yellow

Goshenite—colorless (found in Goshen, Massachusetts)
Heliodor—yellow-green
Morganite—peach, pink, and lavender

Associations Chakras according to color; Moon; yin
See Aquamarine and Emerald entries for their individual associations

Feng shui use Center (spiritual growth); North (personal journey); Southwest (relationships); any direction where healing energy is needed

Bixbite, see Beryl

Bloodstone (Heliotrope)

This form of chalcedony was called the "martyr's stone" in medieval Europe. Legend has it that the green jasper at the foot of the cross was stained with drops of blood from Jesus. Due to the popularity of this story, bloodstone was believed to possess special powers and was a favored stone for carving scenes of the crucifixion. The most famous piece of this genre was created in 1525 by Matteo del Nassaro of Italy. It is titled "The Descent from the Cross."

Bloodstone is also known as *heliotrope.* This name came from the Mediterranean regions where it was said that the stone's coloring was reminiscent of the red glow of the sun *(helios)* setting over the deep green sea. This gemstone was used by the Babylonians for seals and amulets. Bloodstone was believed to render the wearer invisible, an attribute mentioned by Dante in his novel *The Divine Comedy*. Bloodstone is frequently confused with hematite.

A gemstone of courage and strength, bloodstone is useful in helping to remove obstacles. It is associated with honesty and integrity. Bloodstone is believed to help connect with ancestors and support relationships—especially love. It also neutralizes toxins, and attracts good luck and abundance.

Color(s) Green with bright red spots

Associations Aries, Libra, Pisces; root and heart chakras; Earth, Mars; yang

Feng shui use East (ancestors); Southeast (abundance); Southwest (relationships/love); negative directions (remove obstacles/attract luck)

Calcite

This is not a glamorous gemstone, but its use has been important to people in the past as well as the present. Calcite takes its name from the Latin *calcis*, which means "lime." This is not as unusual as it may first seem because limestone is one of the rocks formed from calcite.

Calcite, or calcium carbonate, is one of the most common minerals found on (and in) the Earth, and is one of the most widely collected minerals. Hot springs and other calcium-rich waters leave behind deposits of calcite. Many of Earth's caverns are made wondrous by calcite stalagmites and stalactites.

As was previously mentioned, limestone is formed from calcite; it can contain 50 percent or more of the mineral. Chalk, because of its limestone composition, contains a great deal of calcite. Calcite's more glamorous version is marble, which is simply a recrystallization of calcite.

Calcite has also been called *calcspar. Iceland spar* is sometimes used in place of the name *calcite*, however, it usually refers to its large colorless crystals. Iceland spar has been utilized for prisms in microscopes and other optical instruments. Calcite is employed extensively as an industrial mineral in the production of metals, glass, paint, and rubber, and is the primary component of cement.

Because it is used in a wide range of materials, it is no surprise that calcite is a stone of support, especially for those engaged in the sciences and arts. Calcite also amplifies energy.

Color(s) Colorless and all colors, occasionally multicolored

Associations Cancer; all chakras according to color; Moon, Venus; yin

Feng shui use West (creativity/projects); Northeast (knowledge); North (career); Northwest (benefactors); Southeast (personal resources); negative directions (amplify positive energy)

Carnelian

Carnelian is a form of chalcedony quartz and is found throughout the world. One of the earliest uses of carnelian was in jewelry found in the tomb of Queen Pu-Abi of Sumer, which dated to approximately 3000 B.C.E. It is commonly found in Egyptian tombs, and was apparently as popular as lapis lazuli and turquoise for jewelry. Egyptian myth links carnelian with the goddess Isis, who is purported to have used it to protect the souls of the dead as they made the transition to the afterlife.

Fourth-century Buddhists in China believed in carnelian's protective powers and fashioned it into amulets. Tibetans in the sixth century also used it for amulets, as did Muslims who called it the "Mecca stone"; it was believed that Muhammad wore carnelian in a ring circa 624. This gemstone is also named in the Bible as one of the stones in the breastplate of Aaron.

Ancient Greeks and Romans called carnelian *sardius*. It has also been known as *sadoine* and *pigeon's blood agate*. The word *carnelian* comes from the Latin *carneus*, which means "fleshy." It was probably so named because of the stone's color. Carnelian was popular in Europe during the Renaissance and the nineteenth century.

Belief in carnelian's power of protection remains, as well as its power to calm fears of death. It also protects against anger and soothes grief and sorrow. Carnelian is useful for maintaining calm during times of transition, and is helpful in drawing out a person's talents. This gemstone aids in reaching goals through focus. Carnelian promotes harmony, creativity, and self-worth.

Color(s) Red

Associations Virgo, Aries, Taurus, Cancer, Leo; sacral chakra; Earth, Saturn; yang

Feng shui use Center (harmony); South (success); West (creativity); Southeast (self-worth); negative directions (protection/soothes)

Cat's-eye, see Chrysoberyl

Chalcedony

Chalcedony is a group of microcrystalline quartz gemstones that includes agate, bloodstone, carnelian, chrysoprase, jasper, onyx, sard, and others. It gets its name from the ancient city of Chalcedon in what is now Turkey. The use of chalcedony dates back to the Stone Age where early people employed its durability for weapons, tools, and bowls. It is popular today for jewelry as it was in the classical period of Greece and Rome and nineteenth-century Europe.

For specialized use and associations, refer to individual entries of gemstones.

Chiastolite, see Andalusite

Chrysoberyl

Chrysoberyl is the third hardest gemstone after diamond and corundum (ruby and sapphire). The name comes from the Greek words *chrysos* for "golden" or "yellow" and *beryllos* meaning the variety of stone—beryl. It wasn't until 1789 that it was found to be a mineral separate from beryl.

Chrysoberyl imitators include andalusite, beryl, peridot, spinel, topaz, and zircon. The best known variety of chrysoberyl is the cat's-eye or cymophane. Cymophane is from the Greek *kyma*, meaning "wave," and *phainein*, "to appear," which describes the way the "cat's eye" seems to move. Parallel inclusions in the stone creates the cat's eye effect, however, it must be cut at the correct angle for the effect to be seen. Another dynamically changing type of chrysoberyl is alexandrite. Star chrysoberyls are very rare.

Cat's-eyes have been used for centuries as amulets to attract good luck as well as to protect against bad luck. Cat's-eyes were treasured in first-century Rome, but were not popular in Europe until the late nineteenth century when Princess Louise Margaret's (Prussia) engagement ring brought them into fashion. In Sri Lanka, the "cat's eye" was believed to protect its wearer from evil spirits. Hindus believed it provided protection against poverty.

Cat's-eyes are popular for luck, especially in financial matters. This gemstone fosters optimism and renewal. (*See also* Alexandrite.)

Color(s)	Golden yellow to honey brown and spring apple yellow-green
Associations	Venus; yang
Feng shui use	Southwest (relationships/renewal); Northeast (wealth/luck); use in any area where you want to attract luck; any negative direction (protection/luck)

Chrysolite*, see *Peridot

Chrysoprase

This gemstone derives its name from the Greek words *chrysos*, meaning "golden/yellow," and *prason*, which means "leek." Nowadays we would refer to this yellow-green as apple green.

Used by Egyptians before 3000 B.C.E., chrysoprase is a variety of chalcedony that was also popular during the classical period of Greece and Rome. In an eleventh-century manuscript, Michael Psellius of Byzantine wrote that it improved one's eyesight.[6] Chrysoprase became very popular in fourteenth-century Europe. It was said to be a favorite of Frederick the Great of Prussia, and was used to adorn the Sans-Souci Palace in Potsdam, Germany. This gemstone was also greatly admired by Emperor Charles IV who had it used in the St. Wenceslaus Chapel of the St. Vitus Cathedral in Prague. Its popularity continued into the nineteenth century.

Chrysoprase is attributed with the power for attracting friends, success, and abundance. It lifts emotions and aids in adaptability. This gemstone fosters communication.

Color(s) Yellow-green

Associations Libra; solar plexus and heart chakras; Earth, Venus; yin

Feng shui use South (success); Southwest (relations); Southeast (abundance); Northwest (benefactors) any negative direction (adaptability)

Citrine

This gemstone's name comes from the Latin *citrus* and French *citron* ("lemon"), however, it is anything but a lemon. Citrine is quartz that is yellow due to the presence of ferric iron. Ancient people used citrine for protection, especially against snake bites and evil intentions of others. The earliest use of citrine was in first-century Rome for intaglio (engraved figures or designs). Citrine has been called the "golden stone of wealth" and the "merchant's stone" because of its power to attract wealth. Other names include *Madeira citrine* and *ox blood.*

When citrine forms with amethyst crystals, ametrine is created. Natural citrine is not as common as other types of quartz. Most citrine gemstones are "created" by heating amethyst. Natural citrine is most often a pale yellow. It has sometimes been referred to as topaz quartz, citrine topaz, gold topaz, and Madeira topaz, which is frequently a marketing ploy to pass it off as the more expensive topaz.

Citrine symbolizes joy and aids in getting in touch with one's higher self. As a protector, it raises personal power by helping a person to connect with his or her inner self and tap into hidden strengths. Citrine is also an energizer that aids in emotional healing through awareness. It is useful in emotionally binding families and groups.

Color(s)	Pale yellow, lemon, yellow-brown, orange, dark orange/brown, reddish brown
Associations	Aries, Gemini, Leo, Libra; sacral and solar plexus chakras; Earth, Mercury, Mars, Sun; yang
Feng shui use	Primary gemstone for the Center direction; Center (spiritual growth/guidance); Southwest (relationships); East (community); West (creativity); North (personal journey); Southeast (self-worth/prosperity/wealth); negative directions (protection/strength/healing)

Coral

The coral that is considered a gemstone comes from the species *Corallium rubrum*. The best grows in clear, shallow (ten to forty-five feet deep), warm water. Evidence of its use dates to the Paleolithic period. Its use in Sumer dates to 3000 B.C.E., and it continued to be popular into the classical Greek and Roman era. Greek legend tells that when Medusa died, her drops of blood turned into red coral. In Rome, it was used as a protective amulet for children.

Even today the Italian "horn" luck charm is made of coral. Pliny mentioned a coral trade with India in his writings. Centuries later, Marco Polo wrote about the coral that adorned Tibetan temples. It was also used by Tibetans for mala beads—an aid for prayer and meditation. In twelfth-century England, coral was used as an amulet of protection and an aid during childbirth. Coral was particularly popular in Victorian and Art Deco jewelry.

Coral promotes love and harmony and helps build community. It is useful to clear negative energies and provides protection.

Color(s)	From white to black, most valued are pink and red
Associations	Venus, Neptune; yin
Feng shui use	Southwest (relationships); Center (harmony); East (community); any negative direction (protection/clear negativity)

Cymophane,* see *Chrysoberyl

Dialogite,* see *Rhodochrosite

Diamond

The word *diamond* comes from the Greek *adamas*, which means "invincible" or "I subdue." This is believed to refer to its hardness. Diamond is mentioned in the Bible as one of the twelve gemstones in the breastplate of Aaron. As a symbol of power and protection, diamonds were worn by ancient leaders when they marched into battle. Aristotle and Pliny mentioned diamonds in their writings and made reference to the "valley of diamonds" in India where they had been mined since 800 B.C.E. This was the only known source of diamonds until 1725 when they were discovered in Brazil. In 1866, a new rich source was found in South Africa that set off a diamond rush between 1870 and 1880, which was not unlike the gold rush of the American West.

The ancient Romans used uncut diamonds in jewelry. King Louis XI (1214–1270) of France did not allow women to wear them—not even the queen. Famous diamonds include the Hope diamond which is now in the Smithsonian Museum of Natural History. Its history began in 1669 when it was sold to King Louis XIV of France. The largest cut diamond is the Cullinan which is part of the British crown jewels and is housed in the Tower of London. The first diamond engagement ring was worn by Mary of Burgundy upon her betrothal to Hapsburg emperor Maximilian I in 1477.[7]

The way in which diamonds are formed is as equally fascinating as their beauty. Their story is old (beginning approximately two billion years ago) and their journey long (starting 95 to 120 miles below the surface of the Earth). Diamonds begin as carbon crystals formed by intense heat and pressure below volcanoes. These carbon crystals are transported up to the surface in "pipes" of kimberlite or lamproite rock. If these crystals cool too slowly as they rise to the surface, the result is graphite. The carbon crystals have to rise and cool quickly in order to produce a diamond.

What a disappointment to end up with lead for a pencil instead of a diamond.

The "purest" diamonds are colorless. Colors occur when other substances such as nitrogen (producing a yellow diamond) are present as the diamond forms. A vast array of colorless substances have been used to imitate diamond.

Not only has their beauty attracted people, their hardness (it's *the* hardest mineral) and their ability to conduct heat are attributes that have diamonds employed in a wide range of industrial applications. Diamonds with rich colors such as blue, champagne, green, pink, orange, or yellow are called "fancy" diamonds.

Diamond has been called the "stone of invulnerability" as well as the "king of crystals." In addition to being a symbol of power and wealth, they are also an emblem of love, trust, and commitment. Their powers of protection in battle also extended to protection against disease and pestilence.

The power of diamonds can be utilized to build emotional strength and unite people through reconciliation. In addition, diamonds attract abundance and wealth, and are useful during periods of transformation to help call on inner strength. Diamonds help build relationships and support longevity.

Color(s) Colorless, white, black, and all colors of the spectrum

Associations Aries, Leo, Taurus; all chakras; Venus, Sun; yang

Feng shui use East (family/community); Southeast (wealth/abundance); Southwest (love/relationships); North (personal journey); Longevity direction; any negative direction (good luck amulet)

Emerald

This gemstone's name is said to come from the Greek word *smaragdos*, which among its various meanings includes "green stone."[8] Although this word was applied to all green stones, as far as emerald is concerned, "nothing greens greener" according to ancient scholar and writer Pliny. From Greek to Latin, the name evolved to *esmaraude* in Old French, then *emaraude* in Middle English.

A type of beryl, emerald's deep color is caused by the presence of chromium, which also produces the deep red of rubies. Inclusions (crystal formations within the gemstone) provide depth and a unique identification to each individual stone. On the market, fewer inclusions are more desirable, however, for crystal therapy and feng shui, these features add character and interest. Many emeralds are "oiled" with linseed or cedarwood oil to soften the effects of the inclusions and improve the stone's clarity. The emerald cut—which helps to enhance the color—was created to avoid chipping the corners of the stone.

Emeralds were prized by early civilizations and the Babylonians traded in them as early as 4000 B.C.E. Cleopatra's famed emerald mine was located near Aswan, Egypt. For centuries this mine was thought to be only a legend until it was uncovered in 1818. By then, very few emeralds were found, but the mine did yield tools that were later dated to circa 1300 B.C.E. Emeralds were popular in ancient Egyptian jewelry and many people chose to be buried with them.

Emeralds were also valued by the rulers of India. Shah Jahan, builder of the Taj Mahal, one of the great symbols of love and devotion, is said to have worn emeralds inscribed with sacred texts as a personal talisman. Perhaps his connection with emeralds began this gemstone's link with love. In Europe, emeralds were especially popular from the seventeenth through nineteenth centuries. In the New World, the Aztecs carved emeralds into the shape of flowers and small animals. Emeralds were also used by the Incas and Mayans.

The trapiche form of emerald contains a rare six-spoke pattern around a hexagonal center. *Trapiche* is the name for the Spanish wheel used for pulverizing crops—usually sugar cane. Many other types of green stones, glass, and plastic have been used to imitate emerald.

Emeralds are a symbol of love, and attract good fortune and harmony to all areas of life. This gemstone is also useful against negative energy. It improves memory and some believe it can help a person divine the future—perhaps because it provides access to desires held deeply within. It is associated with spring and rebirth and promotes understanding. (*See also* Beryl.)

Color(s) A wide range of greens

Associations Aries, Cancer, Gemini, Taurus; heart chakra; Jupiter, Venus; yin

Feng shui use Southwest (relationships); Southeast (wealth); North (personal journey/growth); any negative direction (banish negative energy/navigate difficulties)

Feldspar

Feldspar is one of the most common minerals on Earth, and is most widely used industrially. One of its earliest applications was in the clay from which the Chinese made porcelain. The presence of this simple mineral gave the porcelain a fine quality that Europeans could not duplicate for centuries. Feldspar is still used in making pottery, tile, glass, and some plumbing fixtures.

Feldspar gets its name from the Swedish words *feldt*, "field," and *spar*, an Anglo-Saxon name for "easily cleaved minerals." Feldspar's iridescent luster is created by the dispersion of light through its thin layers.

Gem-quality feldspar includes labradorite, sunstone, amazonite, and moonstone. These gemstones are covered separately.

Fluorite

Fluorite's name comes from the Latin word meaning "flow." This is apt for this industrial mineral, which is used in metal processing as flux. The deep blue, banded fluorite found in Derbyshire, England, is nicknamed "Blue John." Decorative objects have been made from this for more than 1,500 years.

Sometimes called the "stone of discernment," fluorite is useful as an aid in finding truth that has been concealed. Fluorite helps to navigate a path of order through chaos, and boosts physical, mental, and spiritual unity through healing. Fluorite nourishes and vitalizes energy while grounding it. This gemstone is also useful to boost the effects of other stones. Fluorite provides strength and protection in times of transition.

Color(s) Clear, black, blue, green, pink, purple, red, yellow; some pieces of fluorite have fluorescent qualities.

Associations Aquarius, Capricorn, Pisces; brow and third-eye chakras; Neptune; yang

Feng shui use South (illumination); North (personal journey); Center (harmony/balance/spiritual healing); any direction that needs a boost of vitality; any negative direction (good luck/protection)

Garnet

The garnet family of complex silicates obtained their name from the Latin word *granum*, which means "grain" or "seed-like." This name most likely evolved from the ancient jewelry that used clusters of tiny red garnets that resembled pomegranate seeds. (Pomegranate is *malum granatum* in Latin.) The garnet group includes almandine, andradite, grossularite, pyrope (now popularly called *rhodolite*), spessartine, and uvarovite.

The use of garnet dates to at least 3100 B.C.E. in Egypt where it was used in jewelry and made into beads. Early mentions of garnet come from the Bible. "Carbuncle" was another name for garnet (as well as "ruby"), which was one of the twelve stones in the breastplate of Aaron. Noah was said to guide the ark at night with a garnet lantern. The almandine variety has been widely used since the classical era of Greece.

For centuries garnets were carried by travelers to protect them from accidents. Ancient Persians considered it a "royal stone" and carved in it images of their kings. In Arizona, tiny granules of this gemstone are called "anthill garnets" because ants push it to the surface while building their tunnels. Garnet saw its first industrial use in 1878 in the United States as a coating for sandpaper.

Like alexandrite, some garnets change color according to the light in which they are viewed. Although rare, there are both four- and six-rayed star garnets. This gemstone occurs in every color but blue. Some garnets are mistakenly called *Arizona ruby*, *Ceylon ruby*, and *Ural emerald*.

Garnet symbolizes faith, devotion, and truth. It strengthens personal power and helps bring victory/success. Garnet aids in the release of kundalini energy and sparks creativity. Garnets tend to have strong supportive energy. This gemstone fosters confidence and success.

Color(s) Almandine—red, orange-red with brown, purple-red
Andradite—black, green, yellow
Demantoid—green, yellow-green
Grossularite—brown, green, orange-yellow, white, yellow
Hessonite—range of orange to brown
Malaya—orange, red-orange, yellow-orange
Melanite—black
Pyrope—red, orange-red, purple-red; all quite dark
Rhodolite—purple, red-purple
Spessartine—orange-brown, red-orange, yellow, yellow-brown
Mandarin—vibrant orange
Tsavorite—bright, dark green
Uvarovite—range of greens

Associations Aquarius, Capricorn, Leo, Virgo; brow and crown chakras; Mars, Pluto; yang

Feng shui use Center (spirituality); North (personal journey/success) Southwest (partnership); Northwest (travel); negative directions (personal power to bring "victory")

Goshenite,* see *Beryl

Heliodor,* see *Beryl

Heliotrope,* see *Bloodstone

Hematite

Hematite gets its name from the Greek *haima* meaning "blood," which is also the root word for *hemoglobin*. This iron oxide's earliest use dates to ancient Europe. Hematite was crushed to produce red ochre, which was used as a pigment to stain figures such as the famous "Goddess of Laussel" (20,000 to 25,000 B.C.E.). This symbolized abundance, fertility, and the life-giving processes of the Great Mother Goddess's blood. Red ochre was also used on burial figurines, as well as the corpse itself. In the ancient Goddess-worshiping cultures, red was the color of rebirth and transformation.

In ancient Egypt, hematite was used for amulets. It was also used to stop hemorrhages. Roman soldiers wore it for protection as they marched into battle. Native Americans use red ochre for ceremonial and war paint.

Hematite is a transformer that turns negative feelings into positive ones, even love. It is a power stone that helps maintain one's sense of self and deflects negativity from other sources. This gemstone enhances memory and balances the body's energy.

Color(s) Metallic/iridescent gray, gray-red, gray-black, brown-red

Associations Aries, Aquarius, Capricorn; third-eye and crown chakras; Mars, Saturn; yang

Feng shui use Center (balance/grounding); Southeast (self-worth); Northeast (self-cultivation); any negative direction (maintain sense of self)

Herkimer Diamond

This gemstone is a type of quartz, but was mistaken for diamond when it was first found in Herkimer, New York, because of its brilliance. Herkimer diamond is sometimes called the "dream crystal" as it aids in getting in touch with one's inner self. It is useful to place under your pillow for dream work to help remember dreams. When utilized this way, it is best to pair it with an amethyst crystal to moderate the energy level. Herkimer diamond is helpful for congeniality among groups of people. It is useful for emotional cleansing because it releases energy blocks. This gemstone also raises energy levels.

Color(s) Clear

Associations Sagittarius; all chakras as a prelude to using other crystals; Uranus

Feng shui use To move and raise energy in any area where energy has stagnated; Northeast (wisdom/self-knowledge); East (community); Southwest (bind relationships); any negative direction (cleansing)

Iolite

This gemstone takes its name from the Greek words *ios*, "violet," and *lithos*, "stone." The Vikings used iolite as a navigation aid; looking through a thin slice of the stone allowed sailors to find the position of the sun in overcast skies.

Iolite is another gemstone with strong pleochroic effects. In iolite's case, three separate colors are displayed, thus making its effect trichroic. It has been mistakenly called *dichroite* (from the Greek *dichrois*, "of two colors"), however, a dichroic effect would consist of only two colors as the name suggests. Other names for iolite include water sapphire and cordierite. The latter is a tribute to P. L. A. Cordier, a French mineralogist of the early seventeenth century.

Iolite's calming energy brings stability to people's emotions. It aids in strengthening faith and promotes cooperation.

Color(s) Blue, violet-blue, gray-blue, green (rare)

Associations Libra, Sagittarius, Taurus; base and sacral chakras; yin

Feng shui use Any direction where stability is warranted; Southeast (wealth/resources); Center (spiritual/calming); Northwest (helpful people); any negative direction (stability/cooperation)

Jade

There are two types of jade: jadeite and nephrite. The name *jade* originated with the Portuguese phrase *piedre de ilharga*, which means "stone of the loins" and explains its use to relieve kidney problems. In Spanish it is *piedra de ijade*, and in French, *piedra de l'ejade*. Nephrite comes from the Greek *nephros*, meaning "kidney." *Lapis nephriticus* is Latin for "stone of the kidney."

Both types of jade are technically rocks made up of microscopic interlocking crystals. Jadeite has a slightly more coarse crystalline structure.

Nephrite is the jade that was used in ancient China as early as 3000 B.C.E. It was a status symbol and was believed to endow powers of immortality. For these reasons it was used in the tombs of emperors and other important people. It was also a symbol of love and virtue. While the ancient Chinese were familiar with jadeite, they did not consider it "real" jade and preferred their nephrite.

In the Americas, jadeite was used by Aztecs, Olmecs, and Mayans for adornment and carvings. In Europe, jadeite axes and tools date back to the Neolithic period. It was also popular through the eighteenth and nineteenth centuries.

Today it is the jadeite that is generally considered the "real" jade. There are three grades of jade: "A" jade is a natural stone with no enhancements; "B" jade is stone that has been treated to diminish any secondary colors; and "C" jade is artificially colored.

A plethora of jade imitators exist. These include aventurine, carnelian, chrysoprase, emerald, garnet, jasper, quartz, glass, plastic, and others.

Jade has been called the "stone of fidelity" because of its connection with love and virtue. It is also called a "dream stone" as it aids in remembering dreams. Like Herkimer diamond, it can be placed under your pillow for dream work. Unlike Herkimer diamond, it does not need a companion stone to soothe the energy. Jade promotes peace and harmony, and is connected with longevity. It helps in finding wisdom to solve problems and bring good luck.

Color(s) Jadeite—black, brown, green, lavender, red, white, yellow
Nephrite—black, brown, green, red, white, yellow

Associations Aries, Gemini, Libra, Taurus; crown chakra; Neptune, Venus; yin

Feng shui use Center (harmony); Southwest (partners/love); Northeast (wisdom); positive directions Life and Longevity; any negative direction (solve problems/bring luck)

Jasper

Jasper was called *jashp* in ancient Persia and *ashpo* in Syria. The Latin name is *jaspis*. Nowadays the names for the many types of jasper correspond with their attributes such as colors or patterns; for example, there is *ribbon jasper* and *picture jasper* (another one of nature's "Rorschach stones").

This gemstone was popular throughout the ancient world for jewelry, bowls, and other objects. Its early use dates to the Paleolithic period. Native Americans employed it for protection when traveling, as well as connecting to the spirits. Red jasper is symbolic of blood and aids in connecting with earth energies.

Jasper is a variety of chalcedony that fosters the ability to nurture. It provides grounding and protection against negativity.

Color(s) Gray-blue, green, orange, red, tan, yellow

Associations General—Leo; all chakras; yang
Red jasper—Taurus; yang
Yellow jasper—Sagittarius; yang

Feng shui use Center (grounding); West (children); Southwest (relationships); East (community/family); any negative direction (protection against negativity)

Jet

This gemstone takes its name from the Old French *jaiet*, which comes from the Latin *gagates* after the town and river Gagas in Asia Minor where the ancient Romans mined it. Technically, jet is not a gemstone since it comes from an organic source. Woody plants that become submerged in bogs and swamps eventually turn into a form of coal. This low-grade coal is called *lignite* and gets its name from the Latin *lignum*, meaning "wood."

When jet is rubbed vigorously with a cloth, it will build up an electric charge and attract small pieces of paper or dust. For this it has been called "black amber" since amber also exhibits this characteristic.

Jet has been used for ornamentation since the Bronze Age. It has been mined in the area of York, England, since approximately 1500 B.C.E. to be used for jewelry. It was also used for adornment by the ancient Romans whose empire extended through York. In fourteenth- and fifteenth-century Spain it was used for carvings, jewelry, and talismans. Through the eighteenth and nineteenth centuries in Europe it was used for jewelry and religious items such as rosaries and crosses. When Queen Victoria went into mourning in 1861, jet became a frequently used gemstone for her black jewelry. In the Americas circa 500 to 1500 C.E., the Aztec, Mayan, Alaskan, and southwest Native American tribes used jet for decoration.

Jet is sometimes confused with obsidian and black tourmaline. Imitators include glass, plastic, and canel (Pennsylvanian anthracite). Black glass in jewelry is referred to as "Paris jet."

In 1213, Arabian botanist Ibn al-Baitar wrote that jet could "drive away venomous beasts."[10] This idea has come down to us in the belief that jet can protect the wearer from illness and banish fear. Jet has a calming influence and lifts one's spirits. It is said to help with the darker side of life. Jet engenders honor and justice and provides protection during times of transition.

Color(s)	Black
Associations	Capricorn; sacral chakra; Pluto, Saturn; yin
Feng shui use	Center (calming/harmony); North (personal journey); Northeast (self-cultivation); South (illumination); any negative direction (to calm and cope/protection)

Kunzite

Kunzite is a form of spodumene. *Spodumenos*, Greek for "burnt ashes," describes the gray-white of many spodumene. Kunzite was named for George Kunz, an early twentieth-century geologist and buyer for Tiffany & Company.

This gemstone is frequently found with morganite and pink tourmaline. In addition to being pleochroic, kunzite is occasionally phosphorescent. This feature and its sensitivity to sunlight—it will fade if exposed to strong light—are the reasons it was dubbed the "evening stone."

Kunzite engenders positive and loving thoughts. It removes negativity, as well as any obstacle that may impede your growth. It provides inner freedom, emotional support, guidance, and protection.

Color(s)	Colorless, green, gray, pink, purple, yellow
Associations	Leo, Scorpio, Taurus; heart chakra; Pluto, Venus; yin
Feng shui use	North (personal journey); Southwest (relationships); Northeast (self-cultivation); any negative direction (remove obstacles)

Kyanite*, see *Andalusite

Labradorite

Labradorite is a type of spectrolite in the plagioclase feldspar family. Its iridescent dispersal of different colors is called *labradorescence.* Frequently found with quartz, this gemstone's most famous deposits are in Labrador, Canada, where it was "discovered" in 1770. Its use by Algonkian tribes in the state of Maine dates to the year 1000.[11] It is sometimes confused with opal.

Labradorite is instrumental in cultivating psychic abilities. With strong powers of transformation, it ushers thoughts from intuition to positive action. It aids in self-reliance and ridding oneself of insecurities. It symbolizes vitality. (*See also* Feldspar.)

Color(s) Light blue, light green, gray, white, pale orange-red, black

Associations Leo, Sagittarius, Scorpio; sacral chakra; Neptune, Pluto, Uranus

Feng shui use Southeast (resources); South (success/reputation); Northeast (self-cultivation); negative directions that occupy the kitchen, dining room, or bedroom (vitality)

Lapis Lazuli

Technically this gemstone is a rock made up of several minerals—mainly lazurite and calcite. Pyrite is also frequently included and gives lapis its gold speckles. Lapis lazuli gets its name from the Latin word for "stone," *lapis*, and the Arabic word for "blue," *azul*.

This gemstone was a favorite throughout the ancient world and has been mined for at least six thousand years. The Mesopotamian capital of Ur had a trade in lapis lazuli that dated back to 3000 B.C.E. It was used extensively in Egypt, Greece, Mesopotamia, Persia, and the Roman Empire. The Egyptians crushed it and used it for a cosmetic. They also employed it for seals as well as carved figurines and vases. They believed that lapis lazuli helped one attain sacred wisdom.

The Europeans called lapis lazuli *ultramarine*, which referred to its blue color and meant "beyond the sea." In Europe it was also crushed and used as pigment for paint well into the nineteenth century. It was used for inlay in furniture, and in St. Petersburg, Russia, it was used to adorn columns in St. Issac's Cathedral and panel a room in Pushkin Palace. Its imitators include synthetic spinel, glass, plastic, and a dyed jasper called "Swiss lapis."

Lapis lazuli is a powerful stone of wisdom that strengthens personal expression and intuition. The awareness it imparts aids in accessing ancient knowledge. It is also a protective stone and promotes tranquility.

Color(s) Blue occasionally with a hint of violet, white, or gold speckles

Associations Aries, Sagittarius; throat and third-eye chakras; Neptune, Venus; yin

Feng shui use Primary stone of the Northeast direction; Center (tranquility/spirituality); Northeast (wisdom/knowledge); North (personal journey); West (creativity); any negative direction (protection)

Lodestone (Magnetite)

This gemstone takes its name from the Latin *magnes*, meaning "magnet." Magnetite is one of two stones that are magnetic. A fable connected with its name tells of a shepherd, Magnes, who accidentally discovers this mineral on Mount Ida (northwest Turkey) when the nails of his shoes cling to the rocks.

Lodestone is one variety of magnetite, which is an iron oxide. The Russian city of Magnitogorsk takes its name from the mineral, as well as its industry as a major iron manufacturer. In the Middle Ages, Polaris became known as the "lodestar" because it attracted the attention of sailors and helped guide them.

Lodestone balances yin and yang energies, provides motivation, and boosts confidence. It aids in finding one's spiritual path.

Color(s)	Black, dark gray, brown-red with black streaks
Associations	Gemini, Virgo; sacral chakra; yin/yang
Feng shui use	North (personal journey); Center (balance/spirituality); South (recognition/fame); any life aspect area where motivation and/or guidance is needed

Magnetite,* see *Lodestone

Malachite

This gemstone is from the carbonate class of minerals and takes its name from the Greek *malaku*, meaning "mallow," a family of herbs. It is a secondary mineral of copper that is created when copper is altered by other chemicals. Azurite, another secondary mineral of copper, is usually found with malachite. It is not unusual to find malachite and azurite banded together in one stone. This is called *azure-malachite*.

Malachite has been used for centuries in jewelry and as household ornaments. Like other gemstones, it was crushed and used as pigment for paint. In czarist Russia it was used to stunning effect to adorn cathedrals and palaces.

Malachite has been called the "stone of transformation." As such, its powers encompass everyday situations, as well as spiritual progression. It is an aid for introspection and balance. This gemstone attracts loyalty and comfort. Malachite is also instrumental in providing protection, promoting success, and banishing negativity. It aids in navigating life's setbacks and challenges.

Color(s)	Light to dark green
Associations	Capricorn, Scorpio; solar plexus chakra; Venus; yin
Feng shui use	Primary stone of the element metal; Center (spiritual progression/balance); Southwest (relationships/loyalty); North (personal journey/introspection); any direction (achieve goals); any negative direction, especially Setback and Difficulty (protection and banish negativity)

Microcline,* see *Amazonite

Moonstone

Moonstone is a type of feldspar that has a shimmering glow. This shimmering effect is caused by the combination of feldspars with different densities and different refractive qualities.

An ancient Roman myth described this gemstone as being created by moonlight. It was used as jewelry by the Romans dating to the year 100 C.E. Many centuries later it was popular in Art Nouveau jewelry. The Romans believed that this gemstone held the image of the goddess Diana. Moonstone was considered sacred in India. There it was believed that if one held it in one's mouth during the full moon, the future could be seen. In sixteenth-century Europe it was believed to help keep a person awake.

Moonstone has been called the "traveler's stone" and believed to provide protection during a journey. This gemstone's association with the moon and the Great Goddess also connects it with mothers and unconditional love. Moonstone encourages inspiration, awareness, and creativity. It also brings good fortune, alleviates fear, and balances yin and yang. (*See also* Feldspar.)

Color(s)	Colorless to gray, blue-gray, brown, green, pink, yellow
Associations	Cancer, Libra, Scorpio; heart chakra; Moon; yin
Feng shui use	Primary stone of the Northwest direction; Center (balance); Northwest (travel); North (personal journey); West (creativity); Southwest (love/mother); Southeast (wealth); any negative direction (good fortune)

Morganite, see *Beryl*

Obsidian

This gemstone is actually natural glass that is created when hot lava is submerged in water. This process forms obsidian's glassy texture. Before the molten rock is cooled, bubbles of air can get trapped between layers, which can produce stunning effects. Some of these have been called *rainbow obsidian* and *sheen obsidian*. When small cristobalite crystals get trapped, *snowflake obsidian* is the result. *Apache tears* is the name given to obsidian that has been worn smooth and round by wind and water, making it perfect for use in feng shui.

Ancient tools and weapons were made with obsidian because of the sharp cutting edge that can be produced by chipping it. Obsidian has also been used for jewelry and mirrors. This gemstone is said to be able to mirror one's soul.[12]

Obsidian is a strong grounding stone and has been called "the protector" for its ability to block negative energy. Popularly used for scrying mirrors, it is believed to aid in divination and seeing into the future by providing insight. Obsidian is useful in dispelling half-truths.

Color(s)	Dark green, dark brown, black Sheens—gold, green, blue, purple, yellow
Associations	Sagittarius; base chakra; Pluto, Saturn; yang
Feng shui use	Primary gemstone for the North direction; North (personal journey); Center (grounding); any negative direction (protection)

Oligoclase, see *Sunstone*

Onyx

This form of chalcedony quartz takes its name from a Greek word that referred to a fingernail, claw, or hoof. It was popular in ancient Greece where a legend tells of Cupid, whether as a form of joke or service, trimmed the nails of the sleeping Venus. Not wanting to see any part of her perish, the Fates turned Venus's fingernail clippings to stone. The Greeks called almost any colored chalcedony stones "onyx."

The Romans limited the name "onyx" to only dark-brown and black forms of chalcedony. They also gave the name "sardonyx" to reddish-brown onyx. (See separate listing for more on sardonyx.) Onyx was also popular during the Renaissance and in the nineteenth century.

Onyx is a stone that provides balance and stability, as well as protection. When used in dream work or meditation, it can help bring guidance and transformation from an inner source. It helps control emotions and negative thoughts.

Color(s)	Black, black-brown with bands of white Sardonyx—reddish-brown, sometimes with white or lighter red bands
Associations	Leo; base and throat chakras; Saturn, Mars; yang
Feng shui use	Center (balance); Northeast (self-cultivation); any direction where you seek transformation; any negative direction (protection)

Opal

This gemstone takes its name from the Sanskrit word *upala*, meaning "precious stone," as well as the Greek word *opallios*, "change of color." In ancient Rome it was called *opalus*.

Containing as much as 10 percent water, the opalescence, or "play of color," is produced by miniscule spheres of silica. Along with the water, these tiny spheres create different rates of light diffraction. The more aligned the silica spheres, the more brilliant the stone's color. This iridescent play of color is also referred to as a stone's "fire." If the stone is heated at high temperatures, the water will be lost and along with it the opalescence. The transparency of the stone and its background color also affect the overall depth of color.

Six-thousand-year-old artifacts found by Louis Leakey are the earliest known use of opal. Ancient Greeks and Romans prized the use of opal and its value was greater than diamond. The Romans nicknamed it "Cupid's stone" because its color can be evocative of a sensuous complexion. The Aztecs also used and valued opal. This gemstone was mentioned in the writings of the ancient scholar Pliny, and again centuries later by Shakespeare in *Twelfth Night*. In the Middle Ages it was called *ohthalmios*, meaning "eye stone," because it was thought to preserve one's eyesight. Also in medieval Europe, if opal was worn by a woman with blonde hair, the opal was believed to keep her hair from turning gray.

During the fourteenth-century plague years, opal became known as a stone of bad luck. It was said that the stones of those who wore it and died changed color. It is now thought that the fairly rapid change of temperature in the wearer's body (from high fever to cold after death) may have been enough to cause an opal to change color. An opal that has a fracture can easily break—a fault that added to this gemstone's connection with bad luck.

Rulers enjoyed opal. It was used in the emperor's crown of the Holy Roman Empire and in the French crown jewels. Opals came to the attention of Queen Victoria when new sources were found in Australia, and she brought it back into popularity during her reign.

Some of opal's imitators are called *opalite* and *Slocum Stone*. Imitators also include glass and plastic.

Opal has been called the "stone of visionaries." The Greeks believed that it had powers of prophecy. The Romans saw it as a symbol of hope. Opal attracts inspiration, insight, and stimulates a wider vision. Some believe it enhances clairvoyant abilities.

Color(s) Black opal—dark to gray-black, dark base color
White opal—milky, light gray, white base color
Gray opal—light gray, gray base color
Fire opal—yellow-red, brown-red, red-orange base color
Hyalite—colorless (from the Greek *hyalos* meaning "glass")

Associations Libra, Scorpio, Sagittarius; throat, third-eye, and crown chakras; Mercury; yin

Feng shui use Primary gemstone for the element water; South (illumination/insight); North (personal journey); Southeast (self-worth); West (creativity); any negative direction (symbol of hope)

Pearl

Pearl is the only gemstone created by a living creature. Its popularity began before the classical Greek and Roman period, and has never since waned. Pearls were used through Asia, the South Seas, and by Native Americans. The Greeks attributed pearls with the ability to help sustain a blissful marriage. Cleopatra wore pearls as did most European royalty centuries later. From the thirteenth to sixteenth century, common people were not permitted to wear pearls as they were seen as being special for only those of royal blood.

Completely natural pearls are rare, and as a result, quite expensive. Most pearls on the market are "cultured," meaning that the process of creating the pearl was initiated by human hands. A pearl begins as an irritant inside an oyster. In nature, this could be a grain of sand; in pearl "farms" it is a bead that is implanted into the mollusk. However it begins, it is not comfortable for the host oyster. In its defense, the oyster releases a secretion called *nacre* to coat the irritant. Nacre is the same material that coats the inside of the oyster's shell and is called *mother-of-pearl.* Any pearl that forms will match the color of the mother-of-pearl. For example, abalone, whose shell is widely used for inlay and jewelry, produces blue-green, green, pink, and yellow pearls. Pearls are produced by both fresh- and saltwater mollusks.

A pearl's luster (reflective quality) adds to its value. The finer the luster, the higher the quality. A pearl's "orient" refers to the iridescent shimmer of its surface. Shape is also important but changes with the times. Various irregular shapes—drops, pears, eggs, nuggets—come in and out of fashion. Perfect roundness has always been highly prized probably because they are difficult to find.

Imitators include coated glass and plastic beads called *Majorca pearls,* which are created by dipping beads into a solution concocted from fish scales.

Pearl symbolizes purity and emotional clarity. It induces emotional balance and openness. Legend says that a pearl is a joyful tear from an angel.

Color(s)	White, cream, silver, gray, gold, blue, green, pink, yellow, black
Associations	Crown chakra; Venus, Moon; yin
Feng shui use	Secondary gemstone for the element water; Northeast (self-cultivation); North (personal journey)

Peridot (Chrysolite)

Peridot, a member of the quartz family, gets its name from the French *péridot*, which originated from the Arabic word for "gem," *faridat*. This gemstone was used by Egyptians as early as 1500 B.C.E. Records of its early mining date to the year 70 C.E. on St. Johns Island in the Red Sea.

This gemstone is born of fire; it is found in rocks from volcanoes. The Hawaiian goddess Pele is said to have shed tears of peridot. Fiery meteors have yielded deposits of peridot, and it has also been found on the moon.

Peridot is a type of olivine. Its darker colors have been mistakenly used for the green demantoid garnet. Peridot's mineral name is *chrysolite*, from the Greek *chrysos*, meaning "gold/yellow." The Greeks also used it for yellow chrysoberyl and other yellowish stones. In ancient Egypt the more yellowish peridot was called "topaz," as it was not distinguished from that gemstone at the time. Some scholars believe that it may have been the "topaz" in the breastplate of Aaron mentioned in the Bible.

Ancient Romans called peridot the "evening emerald" because it maintained its color and did not darken as the sunlight faded. Peridot became popular in Europe after crusaders brought it back from the Mediterranean. It is thought that they initially believed it to be emerald. The "emerald" that graces the Three Holy Kings shrine in the Cologne Cathedral was identified as peridot in the late nineteenth century. It was later mined in Bohemia (the Czech Republic), and in 1900, when additional sources were discovered, peridot became more popular. In addition to decorating churches, it was crushed into a powder and used as a remedy for asthma.

Peridot promotes peace and happiness, and attracts success and good luck. It protects against nightmares and general negativity. Peridot is also instrumental in healing damaged relationships. This gemstone is a symbol of rebirth and renewal. It attracts comfort and builds vitality.

Color(s) Green, green-brown, green-yellow, yellow

Associations Virgo; throat, heart chakra; Mercury, Venus; yin

Feng shui use Primary gemstone for the element fire; South (success); North (personal journey); Southwest (relationships); Northwest (travel); any negative direction (protection); any direction (good luck/vitality)

Quartz

This gemstone takes its name from the old German word *querk-luftertz*, which described the white veins in rocks. Quartz is one of the most common minerals and can be found almost everywhere on Earth. It is the main ingredient in sand found on beaches and in deserts, and it has even been found on the moon. Quartz has been used by humans for thousands of years—at first for tools and weapons, then later for jewelry and decorative objects. Rock crystal objects have been found with human remains in France, Spain, and Switzerland dating to 7500 B.C.E. It was used in ancient Egypt, as well as by the Mayans and Aztecs in the Americas. It was used extensively throughout the fourteenth century for Christian reliquary objects.

Clear quartz has been commonly called *rock crystal.* In Latin it was *crystallus*. In earlier times it was called *crystallos*, from the Greek meaning "frozen," because it was thought to be a permanent form of ice. It's no surprise that the ancients may have believed this, as quartz is always cool to the touch. In our modern world, quartz is widely used in watches and many appliances.

Colorless quartz is common, however, perfectly clear samples large enough to produce crystal balls, bowls, or other objects are not common. Nowadays, glass is frequently substituted. Crystal bowls came from the idea of the Rinn, or "singing," bowl used to create sacred sound in ritual or for meditation. The pure sound it produces is said to balance the energies of all the chakras. Like the chalice and the cauldron, the bowl is symbolic of female energies from which new life emerges.

Smoky quartz is sometimes called *smoky topaz* on the market to garner a higher price. Rutilated quartz is clear quartz with tiny "needles" of rutile arrayed inside. These have been called *Venus hair* and *Cupid's darts.* Tourmalinated quartz is similar, but with dark green or black tourmaline crystals instead of rutile. Sunflower quartz and falcon's eye quartz were popular in classical Greece and

Rome, in Europe during the Renaissance, and in the nineteenth century. The falcon's eye, cat's eye, and tiger's eye effects are created by arrangements of fibrous minerals within the quartz.

Another variety of quartz is a gemstone called *chalcedony*. Instead of a single crystal, chalcedony consists of fine microcrystals. The chalcedony group includes agate, bloodstone, carnelian, chrysoprase, jasper, and onyx.

Quartz is a strong transformer that empowers people and amplifies and focuses energy. It is a protector that also opens the spirit, provides emotional and physical balance, alleviates anger, and reveals distortions.

Color(s) Colorless, white, blue-white, gray-blue, pink, rose, violet, purple, green, brown, yellow

Associations All zodiac signs; all chakras; planet Uranus

Feng shui use White quartz is the primary gemstone for the West direction; quartz is useful anywhere you need to amplify your strengths or where you want to transform a life aspect; Center (balance/spirituality); West (projects/creativity)

Rhodochrosite (Dialogite)

This gemstone takes its name from the Greek *rhodochros*, meaning "rose colored." Although rhodochrosite was "discovered" in Argentina just before World War II, it was used by the Inca as early as the thirteenth century. It is sometimes called "Inca Rose." Rhodochrosite is more widely used as an industrial mineral in the production of alloy steel.

To the opposite extreme, because of its color, rhodochrosite has been called the stone of "love and balance." It is a gentle balancer of emotions and transmits a message of love by drawing in white light. This gemstone engenders love on all levels. It attracts comfort and provides support during times of transition.

Color(s)	Pink (usually pale) with white stripes
Associations	Leo, Scorpio; heart chakra; Mars, Mercury, Venus; yang
Feng shui use	Southwest (love/relationships/partner); Center (balance); Southeast (abundance/comfort); any area that needs love and caring; any area where there's transition (support); any negative direction (offset negativity with love)

Rhodonite

Rhodonite takes its name from the Greek *rhodon*, meaning "rose." This gemstone is pleochroic and has been used for decorative objects—mainly carved beads, boxes, and vases—since the nineteenth century.

Because of its rose color and the associations of roses, rhodonite has been called the "stone of love" and the "stone of brotherly love." This love is on the wider level of caring for humankind and spiritual wealth. Rhodonite brings order to chaotic situations with emotional support, and quells anxiety through clear vision. It balances yin and yang.

Color(s) Dark pinks with veins of black

Associations Taurus; solar plexus and heart chakras; Mars; yang

Feng shui use East (community); Center (balance); Northeast (wisdom); Northwest (helpful people); Southwest (love); any negative direction (banish chaos/calm)

Rose Quartz

This milky rose-pink variety of quartz was cherished in classical Greece and Rome, Renaissance Europe, and the nineteenth century. It was used by the Assyrians (800–600 B.C.E.) for decorative objects, but very little jewelry.

Rose quartz is associated with the heart, love, and beauty. Its calming effects help to balance yin/yang energy. Its warmth heals emotional turmoil and strengthens friendship.

Color(s) Pink, milky rose

Associations Libra, Taurus; heart chakra; Venus

Feng shui use Primary gemstone for the Southwest direction; Southwest (relationships); Center (emotional balance); any direction where emotional healing is needed

Ruby

Ruby takes its name from the Latin word for "red," *rubeus*. It has been called the "king of precious stones," "lord of gems," and "queen of gemstones." This prized stone was written about by the ancients in the Bible, as well as in Sanskrit writings. Ruby is a type of corundum, the crystalline form of aluminum oxide. All other colors of corundum are called *sapphire*. It has been an ongoing debate whether or not pinkish-colored corundum is ruby or sapphire.

In addition to its intense color, ruby can occasionally contain an asterism effect. This appears as a six-pointed star that seems to follow a light source. This star is created by rutile "needles" that align with the ruby's crystal faces.

The intensely colored stones from Burma are called *pigeon's blood*. The ancient Romans considered it the stone of their war god, Mars. Ruby was mentioned in the writings of Pliny and Marco Polo. It is believed that a ruby that turns a darker color indicates bad luck will befall the owner. Henry VIII's first wife, Catherine of Aragon, is said to have expected difficulties because of her darkening ruby.[13]

Because ruby is rare, almost any deep-red gemstone is used as an imitator. The so-called balas ruby is spinel, and Brazilian ruby is actually pink topaz. Nongem quality rubies are used as movement bearings in watches.

For centuries ruby has been a symbol of devotion and love. In addition, it is believed to attract wealth, inspire wisdom, and strengthen self-esteem. This gemstone engenders loyalty and generosity. It dispels fear and protects from all forms of negativity.

Color(s)	Red, brown-red, pink-red, purple-red
Associations	Cancer, Leo, Scorpio; heart chakra; Mars, Sun; yang
Feng shui use	Northwest (father/assisting people); Southeast (self-worth/wealth); Southwest (relationships/loyalty); Northeast (wisdom); any negative direction (protection)

Sapphire

This aluminum oxide corundum comes in almost every color except red. As is mentioned under ruby's entry, when it is red it is called a ruby. In ancient India and southeast Asia, sapphires were thought to be "unripe" rubies. In Latin it is *sapphiru*, and in Greek *sappheiros*—both refer to the color blue. The Sanskrit *saripruja*, which may have been closer to the original word for "sapphire" was also applied to lazurite and lapis lazuli.

Sapphire's long history has connected it with the heavens and the heavenly realms. An ancient Persian legend told of the Earth balancing on a huge brilliant sapphire whose reflection gave its color to the sky. Another legend tells of the Ten Commandments being delivered to Moses on tablets of sapphire. For these reasons, rulers of both church and state have used sapphires as emblems of their sincerity and wisdom.

Sapphire was used by Etruscans, Egyptians, Greeks, and Romans. It was used frequently in medieval Europe by royalty in brooches and rings. The British crown jewels sport a number of sapphires. This gemstone has been extremely popular since the eighteenth century.

In addition to its use for adornment, sapphire was utilized in the seventeenth and eighteenth centuries as one of a number of mineral ingredients in the varnish of Stradivarius and Guarneri violins. Some believe that the use of crystal particles is what gives these instruments their rich sound.

This gemstone is particularly known for its star variety. This effect is caused by tiny "needles" of rutile that align within the stone's planes. Centuries ago these stars were used as amulets and "guiding stars" by travelers. These were also called the "stone of destiny" with its three lines symbolizing faith, hope, and destiny. The six-ray star is somewhat common, but twelve-ray stars are rare. Also rare is the sapphire cat's eye. Some sapphires change color from blue in natural light to violet. Another rare type of sapphire is the *padparadscha.* This reddish-yellow-orange sapphire was named for the color of the lotus flower. The earliest source for these was in Sri Lanka near Ratnapura, whose name meant "city of gems."

While sapphire comes in a wide range of colors, the most popular is blue. Many of these are heated to remove unwanted trace elements that cause a stone to look cloudy, too dark, or too light. Sapphires in colors other than blue are sometimes called "fancy" sapphires.

Sapphire is a gemstone of prophetic wisdom (gained through mental clarity and intuition) that helps one find his or her purpose in life. It also provides healing energy.

Color(s)	Blue, green, pink, purple, violet, orange, yellow, black, colorless
Associations	Aquarius, Libra, Virgo; third-eye and crown chakras; Neptune, Moon; yin
Feng shui use	Northwest (travel); North (personal journey); Northeast (wisdom); Southwest (love/devotion); anywhere healing energy is needed

Sard

This gemstone takes its name from the Greek *sard*, meaning "reddish-brown." It is a type of chalcedony that, if a little more red, would be carnelian.

Sard was used by the Mycenaeans (1450–1100 B.C.E.) and the Assyrians (1400–600 B.C.E.). Along with carnelian, sard was used for engravings and seals by the ancient Romans.

In the fourth century, sard was used to heal wounds. It is a strong protector against negativity and boosts self-confidence.

Color(s)	Rich reddish-browns
Associations	Mars; yang
Feng shui use	South (respect/reputation); Southeast (self-worth); Southwest (relationships); any negative direction (protection); a general healer

Sardonyx

This banded form of cryptocrystalline quartz takes its name from the Greek *sard*, meaning "reddish-brown," and *onyx*, Latin meaning "veined gem." In ancient Egypt (2000 B.C.E.), sardonyx became popular because it was commonly available to most people, whereas precious gemstones were accessible only to royalty and upper classes. It was mentioned as one of the stones in the breastplate of Aaron. Jews frequently wore it and used it to adorn their temples.

Cameos of sardonyx became popular in classical Greece and Rome and continued to be for many centuries. It was also frequently used as the stone at the end of the chain in the fob watch. Napoleon is said to have worn a carved sardonyx from Egypt on his watch chain.[14]

In the Middle Ages, sardonyx was used for healing, especially the eyes. The coolness of the stone when placed on the eyelids was said to bring respite from discomfort. During the Renaissance its power for aiding communication made it a favorite for speakers. It was also a symbol of a happy marriage.

A cameo of Queen Elizabeth I of England was carved in sardonyx and given by her in a ring to the earl of Essex. This is the ring with which she also pledged her help to him. Years later when the earl was convicted of treason and sentenced to death, the ring fell into the wrong hands and did not get back to the queen in time to save him.

Sardonyx aids in clear, focused thinking, and is used to enhance communication between partners.

Color(s) Red-brown

Associations Mars; yang

Feng shui use Southwest (partner/love); West (projects); Northeast (wisdom)

Serpentine

This gemstone takes its name from the Latin *serpens*, meaning "snake," which refers to its snakeskin-like patterns and coloring. Since ancient times it was worn as an amulet to protect the wearer from snakebites. It has been used for decorative objects and is a popular stone for African sculptors in Zimbabwe. Serpentine has been utilized as a substitute for jade and is called *Korean jade* or *immature jade*.

This gemstone's connection with serpents extends to its ability to draw up kundalini energy. It augments meditation.

Color(s)	Green, brown-green, back-green, brown, yellow
Associations	Gemini; heart chakra; Saturn; yang
Feng shui use	North (personal journey); Center (balance/ spirituality)

Sodalite

Sodalite is a rock-forming mineral that is frequently confused with and used as a substitute for lapis lazuli. Its name may have come from the Latin *sodanum*, meaning "a cure for headaches." (*Suda* in Arabic means "headache.") Since the seventeenth century it has been used for jewelry.

Sodalite enhances community relationships and aids in resolving issues logically. It helps clarify purpose and direction in life. This gemstone is supportive for meditation and the pursuit of wisdom.

Color(s)	Blue, lavender-blue, green, gray, white, and colorless
Associations	Venus; yin
Feng shui use	East (community); Southwest (relationships); Northeast (wisdom); North (career); South (success); any negative direction (resolve issues)

Sphene

This gemstone takes its name from the Greek *sphen*, meaning "wedge," which refers to its wedge-shaped crystals. Sometimes called *titanite*, sphene is an ore of titanium and is used industrially in the manufacture of airplanes. It is confused with topaz, yellow beryl, and, because it can be pleochroic, chrysoberyl.

Sphene promotes intellectual and spiritual endeavors.

Color(s)	Green, black, brown, yellow, white
Associations	Mercury; yang
Feng shui use	Center (spirituality); North (foundation); North-east (knowledge); any negative direction where you need to think issues through

Spinel

This gemstone is said to take its name from the Latin *spinella*, "little thorn," as well as *spina*, "spine" or "thorn." Either way, it is most probably because of spinel's pointed octahedral shape. Spinel is a magnesium aluminum oxide and is frequently found with rubies and sapphires (aluminum oxide corundums). Spinel has also been called "balas ruby," which was a name generally applied to gemstones that were borderline red/red-violet.

In ancient Sanskrit writing, spinel is referred to as the "daughter of ruby." As it turns out, many famous rubies have been identified as spinel. The Timur ruby, which has been traced back to fourteenth-century India, is now among the British crown jewels. The Black Prince ruby—named for Edward (1330–1376), son of King Edward III of England—had been given to him in 1367 by the king of Castile.[15] This spinel now has a place in the Imperial State Crown and is housed with the other crown jewels in the Tower of London. The ruby in the crown of Russia's Catherine II (1762) is a spinel. Spinel has been popular in classical Greece and Rome, during the Renaissance, and in the eighteenth and nineteenth centuries.

Color-changing spinels have been found. These change from blue in natural light to purple in artificial light. A green variety of spinel is sometimes called *chlorospinel. Ceylonite* is a black variety of spinel that is also called *pleonaste.*

Spinel is a gemstone to have during difficulties as it is a general healer that helps reconcile differences and relieve sorrows. It is also a stone of protection and can aid in attracting wealth. It enhances one's ability to overcome obstacles and setbacks.

Color(s) Red, blue, purple, pink, violet, orange; colorless, green, and yellow are rare

Associations Pluto; yang

Feng shui use Southwest (relationships); Southeast (wealth); any negative direction (protection, or where you need to deal with sorrow or make amends with others); Setback (overcome obstacles)

Staurolite

This gemstone takes its name from the Greek *stauros*, meaning "cross." The "twinning" of its crystals frequently occurs at right angles to create the shape of a cross. It has also been called the "fairy cross" because, according to legend, these crystals were formed from the tears shed by fairies upon receiving the news that Jesus had been crucified. Staurolite crosses, like those of andalusite, were worn as amulets by Christian pilgrims.

Staurolite is a stone of protection and good luck. It is also good for grounding energy.

Color(s) Red-brown, yellow-brown

Associations Pisces; crown chakra; yin

Feng shui use Center (grounding); Northwest (travel); any negative direction (protection); any area where you want to bring luck

Sunstone (Oligoclase)

This gemstone is a type of feldspar that glitters because of inclusions of hematite and/or goethite. While the most common colors are like those of the sun, a green sunstone is commonly called *aventurine feldspar.* Sunstone has been used by people in ancient India and Greece, as well as Native Americans in Canada for rituals of sun healing and connecting with spirit guides.

Sunstone relieves stress and banishes fear. It is also useful when working with the spiritual realm. (*See also* Feldspar.)

Color(s) Yellow, orange, red, brown, pink, peach, green, and gray

Associations Libra, Leo; sacral and solar plexus chakras; Sun; yang

Feng shui use Center (spirit); North (personal journey); Northeast (self-cultivation); any negative direction (deal with fears/relieve stress)

Tanzanite

Tanzanite is fairly new on the scene, having been discovered in 1967 while a prospector was searching for sapphire. This gemstone is only found in Tanzania, not far from Mount Kilimanjaro. In the 1970s it was only available through Tiffany & Company.

Tanzanite is a trichroic gemstone, meaning that it will appear as one of three colors when viewed from different directions. These colors are most frequently blue, purple, and bronze-brown. Straight from the earth the predominant color is brown, however, heat treatment coaxes out the "velvety-blue."

Tanzanite is a type of zoisite—a mineral named for Baron Siegmund Zois, an Austrian scholar (1747–1819). Tanzanite aids in dealing with change and weathering difficulties.

Color(s) Blue, violet, purple, bronze-brown
Chrome tanzanite—green

Associations yang

Feng shui use Northwest (benefactors); Southeast (personal resources); Northeast (knowledge); Difficulty (deal with change)

Topaz

There are two possible origins for this gemstone's name. One is from the Sanskrit word *tapas,* meaning "fire." The other is the Greek name for an island in the Red Sea, Topazion. The island was frequently shrouded in mist, so its name came to be synonymous with seeking. This may be the source of the belief that topaz could help one be clear-sighted—not only physically to correct eyesight, but also to "see" one's way through problems. Topaz was also endowed with the ability to make the wearer invisible (as though the mists of Topazion Island could be summoned). In the Middle Ages this cloaking capacity of topaz was attributed to its ability to call forth guardian angels in time of emergencies. At that time it was also believed to cure fevers and ease childbirth.

Because of its most prevalent color, topaz was associated with the sun god Ra in ancient Egypt and Jupiter in ancient Rome. In the classical era of Greece and Rome, the name "topaz" was used for most yellowish stones. Its popularity grew in the thirteenth century and it has remained strong ever since. The tag name "Imperial" topaz for the deep-pink and orange-red stones originated because of its renown in the eighteenth and nineteenth century with Russian czars and czarinas. By the mid-nineteenth century it was highly prized and expensive. The large colorless topaz in the Portuguese crown was originally thought to be a diamond when it was found in 1740.

Blue topaz, which has been growing in popularity, is usually created by irradiating pale, white, or colorless stones.

Topaz helps one focus on what one wants to achieve. Called the "stone of the sun," topaz brings warmth and light as well as healing to those who need it. Topaz attracts abundance and love. It also is instrumental in getting energy moving. This gemstone alleviates tensions and promotes communication.

Color(s)	Yellow, yellow-brown, orange-brown, red, blue, green, violet
Associations	Sagittarius, Scorpio, Taurus; solar plexus chakra and up; Mercury, Sun; yang
Feng shui use	Southeast (net worth/abundance); Southwest (love/partner); any area where energy has stagnated; anywhere to alleviate tension

Tourmaline

Tourmaline is a large group of minerals that come in all colors of the rainbow, as well as colorless. This gemstone gets its name from the Sinhalese (Sri Lanka) words *turamali* ("mixed-colored stones") and *toramalle* ("something little of the earth"). These terms were first used in reference to green, brown, and yellow stones and mainly zircon.

Tourmaline has been utilized for its beauty for several thousand years. A tourmaline intaglio of Alexander the Great dating to between 300–200 B.C.E. is on display in the Ashmolean Museum in Oxford, England. While Nordic jewelry with tourmaline has been dated to the year 1000, it was not until 1703 that Dutch traders brought it into widespread use in Europe from Ceylon. Many of the rubies in seventeenth-century Russian crown jewels have turned out to be tourmaline.

In Victorian England the black variety of tourmaline called *schorl* was popularly used for mourning jewelry. The word *schorl* is an old mining phrase that meant "unwanted material." Tourmaline was the favorite of famous gemologist George F. Kunz, who gathered it for a number of collectors, including museums. He also introduced it to Tiffany & Company.

This gemstone is piezoelectric and pyroelectric. Most tourmaline is pleochroic and it is not unusual to find a tourmaline crystal that is half one color and half another. Tourmaline's wide variety of colors come from a range of chemical compounds. Few of the type names are used today in favor of simply identifying stones by their color or pattern. For example, watermelon tourmaline is pink, white, and green like a slice of the fruit. A few of the names still in use include *rubellite* (from the Latin word for "reddish of color") and *elbaite*, which is the tourmaline that comes from the island of Elba. The latter is the type most frequently utilized in jewelry. Paraiba tourmaline simply describes the area of Brazil from which it comes. Paraiba tends to come in very brilliant blues and greens.

Tourmaline is attributed with healing powers and the ability to neutralize negative energy. It can help to provide insight and attract inspiration. It is associated with compassion and meditation. Tourmaline aids in handling grief. It dispels fear for positive change.

Color(s)	Colorless, blue, black, green, lilac, violet, brown, pink, red, orange, yellow
Associations	Libra; all chakras according to color; Venus, Saturn, Pluto; yin
Feng shui use	North (personal journey/insight); West (creativity/inspiration); Northeast (wisdom); Southwest (partner); any direction according to color; any negative direction (neutralize negative energy/protection)

Tsavorite

This gemstone is a grossular variety of garnet, which was only discovered in Kenya and Tanzania in the late 1960s. It takes its name from the Tsavo game preserve in Kenya. Brilliantly emerald-like in color, tsavorite was brought to world notice by Tiffany & Company. It is usually found with a coating of quartz or scapolite, and was originally thought to be demantoid, the only other type of green garnet.

Use tsavorite to vibrate with the heart chakra.

Color(s) Light to emerald green, yellow
Associations Heart chakra; Mars, Pluto; yin
Feng shui use Center (balance/harmony)

Turquoise

This gemstone takes it name from the Middle Ages and an Old French word *turqueise*, which meant "Turkish"; stones arrived in Europe from the Middle East. Its more ancient name was *callais*, from the Greek words *kallos lithos*, "beautiful stone." Turquoise is one of the most widely used gemstones. It comes from dry (arid and semi-arid) regions and is a secondary mineral in copper deposits.

The use of this gemstone can be traced back to 5500 B.C.E. in Egypt where turquoise was found in the tomb of a queen. It was also used for amulets and ground into powder for cosmetics. Persia, where it was used for religious carvings, was the ancient source for fine turquoise. Turquoise is one of the twelve mentioned in the Bible on the breastplate of Aaron.

In China, the use of turquoise began before the year 1000. Hindus and Tibetans utilized it and ascribed it with the ability to bring good luck. In parts of the Middle East, verses from the Koran were engraved onto tablets of turquoise. It was not used in Japan until the eighteenth century. In the Americas, turquoise was mined in New Mexico since the fifth century and used by the Anasazi, Apache, Navajo, and Zuni. It was mined by the Aztecs since approximately 900. Some Southwest American tribes carved turquoise for beads, and others utilized it for currency in trade with Mexico. Warriors tied turquoise to their bows to aid in making precise shots.

While the use of turquoise dates to approximately 500 B.C.E. in eastern Europe (Siberia), it did not become popular in western Europe until the late Middle Ages. Resembling the color of the forget-me-not flower, turquoise jewelry was given for remembrance and affection. Such use was documented by Shakespeare in *The Merchant of Venice* with the ring given to Shylock by Leah.

Howlite, dyed chalcedony, glass, and plastic are used to imitate turquoise. Synthetic turquoise is sometimes called *neo-turquoise* or *neolite*.

Since the days of ancient Persia, turquoise has been a stone of good luck. During the Middle Ages in Europe it was believed to protect both horse and rider from danger. It was also thought to change color if the wearer was ill. This may actually happen because turquoise is slightly porous and will react to oils, soaps, and perspiration. It is also sensitive to strong sunlight.

Many cultures from the Middle East to the Americas considered this gemstone a symbol of the sky. The Apaches believed that it contained the powers of the heavens and the sea. It is still considered to be a "bridge" between heaven and earth, and is used for spiritual cleansing. Turquoise has been called a "stone of communication" and provides a balance of yin/yang energy. It protects against negativity, and can attract love. Turquoise is a general healer.

Color(s) Blue, blue-green

Associations Aquarius, Sagittarius, Taurus; throat chakra; Venus, Neptune, Moon; yin

Feng shui use Center (balance/spiritual cleansing); North (personal journey); Southwest (partner/love); any negative direction (protection)

Zircon

This gemstone takes its name from the Persian word *zargun*, which was a compound of *zar*, "gold," and *gun*, "color." It was well-known in India, and was mentioned in the Hindu legend of the Kalpa Tree. This tree, which was given to the gods as a gift, was laden with fruit of various gemstones. Its leaves were made of zircon. This gemstone was popularly used by the Assyrians. During the classical period of Rome and the Middle Ages, yellow zircon was favored.

Zircon did not become popular again in Europe until the 1920s. The colorless zircon is so brilliant that it was used widely as a substitute for diamond. These have sometimes been called *Matura diamonds.* Zircon is frequently confused with zirconia, which is a lab-grown diamond imitation. Types of zircon include hyacinth and jacinth, both of which were mentioned in the Bible. The blue variety, called *starlight zircon*, is created by heating yellow zircon.

Since the Middle Ages, zircon has been believed to aid in spiritual growth and to promote prosperity and wisdom. This gemstone is also instrumental in finding beauty and peace.

Color(s) Colorless, blue, green, brown, orange, red, yellow

Associations Scorpio, Taurus; crown chakra; Sun; yang

Feng shui use Center (harmony/spiritual growth); North (personal journey); Northeast (wisdom); Southeast (abundance/prosperity)

1. Baur and Boušk, *A Guide in Color*, 148.
2. Mithridates the Great (died 63 B.C.E.) was king of Pontus, a small kingdom in Asia Minor and now part of Turkey.
3. The epic work of the Roman poet Ovid (born 43 B.C.E.) encompasses the history of the world.
4. Baur and Boušk, *A Guide in Color*, 102.
5. Morgan donated his collection to the American Museum of Natural History in New York.
6. Sofianides and Harlow, *Gems & Crystals*, 158.
7. Ibid., 38.
8. Ibid., 62.
9. Baur and Boušk, *A Guide in Color*, 166.
10. Sofianides and Harlow, *Gems & Crystals*, 187.
11. Ibid., 128.
12. Andrews, *Crystal Balls*, 56.
13. Sofianides and Harlow, *Gems & Crystals*, 51.
14. Heaps, *Birthstones*, 83.
15. Prince Edward became known as the Black Prince, but it is not certain that this was because of the color of his armor or whether it referred to his temper.

Appendix A

How to Buy Crystals and Gemstones

There is no right or wrong way to buy a crystal or gemstone. The only criterion is to approach the acquisition of these stones with an open mind and heart. Be sure that you are in a good frame of mind. If you are having a bad day or are upset with some event in your life, you may tend to be less open to receive vibrations from the stones because your energy field may be in turmoil. If you are planning an excursion specifically to buy stones for feng shui use, take time before leaving home to meditate or at least ponder the purpose for the stones you seek.

Plan how you are going to use the stones. Over time you may want to build up a collection with sets of stones for different purposes because the particular energies that are amplified in one use may not be what you want to enhance in another use. It's a good idea to write up a shopping list for your stone collecting. If you are like me, you'll find that it is easy to get sidetracked like a kid in a candy store when confronted by an array of beautiful crystals and gemstones.

When aligning your chakra energies, the color of stone is more important than the type. For this work you will want to go for the most vibrant tones you can find. The colors are red, orange, yellow, green, blue, indigo, and purple or violet. For crystal therapy, coloring is almost irrelevant, but you may want to gather types of stones that contain healing energies such as agate, amber, amethyst, calcite, garnet, jasper, lapis, peridot, sodalite, and turquoise. If you plan to use them in a laying-on of stones for sky-to-earth/earth-to-sky directing of energy, look for stones that have a point (this is easy with many crystals) or are somewhat pointed. A stone that is angular may have one point that is more pronounced than the others. Assess it by asking yourself whether or not you think energy would be apt to flow in the direction of that point. You may also want to acquire a particular type of stone for a particular healing situation. In this case, use the guide in this book or other books that provide details on the specific healing properties of crystals and gemstones.

Stones used specifically for elemental energy work can be selected by color or type or both. Refer to Table 2.2 on page 20, or Table 6.1 on page 71, depending on whether you are working with Western or Chinese elemental energy flow. Because there are different elements in these two systems and their energies interact with each other differently, it is best to acquire separate sets of stones if you intend to work with both types of elemental energy flow.

When working with crystals and gemstones in feng shui, there are three methods for utilizing the stones. Each method requires a different way of selecting stones. The simplest method is to employ a stone according to the color associated with each element. For example, when working with the element fire, a red stone such as garnet would be used. For metal, a gray or white stone such as white quartz or moonstone would be used. An alternative to selecting stones by elemental color is to choose them according to elemental shape. The shapes are based on the energy movement associated with each element. For example, the energy movement of fire is upward or rising and the associated shape is a triangle. If you can find a red crystal or gemstone, or a piece of peridot or obsidian that is triangular, you will increase the energy level of that element because you will be calling on two of its associated attributes. With the example of metal, its energy moves inward and its shape is round. A round piece of malachite or azurite, or a white or gray stone, will enhance the level of metal energy. Refer to Table 6.2 on page 82 for the full listing of associated energy movement.

The second method of selecting stones calls on components of the stones that associate them with the elements. In the case of fire, peridot or obsidian would be used because both are born of volcanic action. In the case of peridot, it could also come from a fiery meteor. For the element metal, malachite or azurite can be used because they contain more than 50 percent of the metal copper.

For the third method, decide whether you want to work with gemstones designated by birth month or by signs of the zodiac. If using gemstones by birth month, you may find that stones on an older or national list may resonate with you more than the modern jeweler's lists. Go with what your heart tells you is right. You can use multiple stones by selecting others according to the day of the week and hour of the day you were born, as well as a stone that represents an angel. When you want to affect a particular aspect of your life or heal an ailment, select a particular type of stone with the relevant associated energy. For example, if you are working on your relationship with your mother, mother (and relationships in general) is associated with the direction Southwest. The color for this direction is pink. Rose quartz can be used here because it is pink and it is associated with strengthening relationships. Beryl could also be employed because one of its colors is pink and it stimulates communication and acceptance.

If you use gemstones for healing and you have a headache for example, lie down in the Northwest (associated with the head) sector of your room and place an amethyst, aquamarine, turquoise, or blue tourmaline on your forehead. I would also suggest darkening the room and using a light treatment of lavender aromatherapy. As in previously mentioned methods of utilizing gemstone energy, you may want to look for stones with the elemental shape appropriate to the direction in which they will be employed to amplify the energy.

As you begin to assemble your various sets of gemstones, it's a good idea to match the size of the stones within each set as closely as possible to maintain balance. When this is not possible—for example, peridot is usually only readily available in small pieces—use several small stones with an aggregate size that seems appropriate to you. Acquiring stones is a personal matter and you must depend on your own intuition and reactions to individual stones. If two pieces of peridot look and feel equal to the others in a set, use two. If three pieces are right for you, go with it.

When buying gemstones for jewelry, you most likely want to find ones with as few imperfections as possible. For energy work, the opposite is often the case. Stones with inclusions, frost, or minor flaws that in jewelry might be a detraction actually reveal more of their true character and tend to release energy more readily.

There are many sources for crystals and gemstones. Nowadays, a common question is, "Is it okay to buy them on the Internet?" Like everything else in acquiring your stones, it's up to you to decide what is right for you. I have never bought them through a website because I prefer and usually recommend that a stone be viewed and held in the hand. In this way, you can immediately interact with a stone's energy and determine if it's right for you. Buying in person also allows you to see the size and shape of the stone, which you cannot do over the Internet. That said, if you have no local source for crystals and gemstones, or if you cannot find a particular stone in your area, then by all means try the Internet. Since most websites also offer buying over the phone, you may be able to speak with someone to describe the size and shape you want. Also check the vendor's return policy in case you don't get what you're looking for.

Crystals and gemstones are cropping up in many types of stores. New Age shops and stores that cater to Wiccans and Pagans are excellent sources. You may also find them in novelty shops (use caution as these may not be labeled properly or even be real stones), popular earth science stores, and specialty shops for gardeners who like to attract wildlife into their yards. I have acquired a number of stones at my local "birding" (birdfeeder) shop. Museum gift shops are another fine place to locate gemstones. You may also be surprised to look in your local Yellow Pages under "Rock Shops." Rock and fossil shops are great sources and you can count on them to correctly identify the less-popular types of stones. Ask and look for notices of upcoming rock/fossil/gemstone fairs, shows, or swaps. These are excellent venues for finding a stunning range of crystals and gemstones, as well as learning more about them.

Once you have found a place to buy your crystals and gemstones, and have located a stone that is on your shopping list, take a few moments to hold it in the palm of your hand. Close your fingers over it or hold it between both palms and sense its energy. Think of how you are planning to use it. Are you going to emphasize an element with it? If you are going to use it for water, for example, think of water. If the stone still feels good as you think of an element, then it's the right stone. Don't expect a thunderbolt or choir of angels to let you know you've chosen the appropriate one. Clues tend to be subtle, or you just might "know" it's the right stone for you. Also listen to yourself if something doesn't "feel" right. You may not be able to pinpoint what it is that makes you react or feel a certain way with a particular stone, but if this happens it's just not the right one with which you should do energy work. Don't take it personally, just listen to your heart.

Stones that you find or trade or receive from friends can be very special. Stones that you find are gifts directly to you from Mother Earth. If you are not sure about the identification of a stone you find, take it to a rock shop or show and ask for more expert advice. Don't be shy, because most rock hounds would be honored to share their knowledge. When you exchange stones with friends, you will know a little bit of the history of that individual stone, which can make it more special for you.

Whatever way a stone comes into your life, follow the suggestions in chapter 2 for cleaning and preparing it for energy work. Even one that comes from your best friend should be prepared before use in order for it to function more richly and fully with your own energy.

Appendix B

Myths and Misleading Names

Like the proverbial old wife's tale, some common advice about gemstones is not true. This includes: all sapphires are blue; the darker the color, the better the quality of gem; real diamonds cannot break; and, if it's old it's probably real. Occasionally the name of a stone—such as sapphirine or emeraldine—is created in order to associate it with a particular gemstone. The largest number of imposters include: diakon, diamanite, diamite, diamon-brite, diamonette, diamonte, diamontina, diamondite, and diamothyst, all of which are passed off as diamonds. The following chart provides a brief listing of misleading names.

Misleading Name	Actual Stone
Alexandrine	Synthetic alexandrite
American ruby	Usually rose quartz
Balas ruby	Spinel
Beach moonstone	Quartz
Black amber	Jet
Blue malachite	Azurite
Blue moonstone	Chalcedony dyed blue
Brazilian ruby	Pink topaz
California moonstone	Chalcedony
Cape May diamond	Quartz
Ceylon opal	Moonstone

Misleading Name	**Actual Stone**
Ceylon ruby	Garnet
Chinese turquoise	Usually quartz or soapstone
Colorado topaz	Citrine
Desert amethyst	Glass
Emeraldine	Dyed chalcedony
Emeraldite	Tourmaline
Golden topaz	Honey-colored quartz or citrine
Green onyx	Dyed chalcedony
Hot Springs diamond	Quartz
Immature jade	Serpentine
Korean jade	Serpentine
Madeira topaz	Citrine
Majorca pearl	Coated bead
Manchurian jade	Soapstone
Matura diamond	Zircon
Montana ruby	Garnet
Neolite	Synthetic turquoise
Nevada black diamond	Obsidian
Orange topaz	Quartz
Paris jet	Glass
San Diego ruby	Tourmaline
Sapphirine	Chalcedony or glass
Smoky citrine	Quartz
Smoky topaz	Quartz
Swiss jade	Dyed jasper
Swiss lapis	Dyed jasper
Vienna turquoise	Glass
Water sapphire	Iolite

Appendix C

Directional Listing of Gemstones for Feng Shui

The following list has been compiled from Part II to provide quick access by direction for the application of gemstone energy in feng shui.

Center

Agate—balance, connection to the natural world
Alexandrite—spirituality
Amber—balance, calming
Amethyst—balance and harmony, spiritual development
Ametrine—balance, spiritual development
Andalusite—grounding, balance, spiritual growth
Azurite—spiritual growth and guidance
Beryl—spiritual growth, healing
Carnelian—harmony
Citrine—spiritual growth and guidance
Coral—harmony
Fluorite—harmony, balance, spiritual healing
Garnet—spirituality (faith/devotion)
Hematite—balance, grounding
Iolite—spiritual growth, calming
Jade—harmony
Jasper—grounding
Jet—calming/harmony
Lapis lazuli—tranquility/harmony, spirituality

Lodestone—balance, spirituality
Malachite—spiritual progression, balance
Moonstone—balance
Obsidian—grounding
Onyx—balance
Quartz—balance (emotional and physical), spirituality
Rhodochrosite—balance (white light)
Rhodonite—balance (through wider vision)
Rose quartz—balance (emotional)
Serpentine—balance (kundalini), spirituality
Sphene—spiritual work
Staurolite—grounding
Sunstone—spiritual matters
Tsavorite—harmony, balance
Turquoise—balance, spiritual cleansing
Zircon—harmony, spiritual growth

North

Andalusite—success
Aventurine—career success
Beryl—personal journey
Calcite—career
Citrine—personal journey
Diamond—personal journey
Emerald—personal journey, growth
Fluorite—find path for personal journey
Garnet—personal journey, success
Jet—personal journey
Kunzite—personal journey (inner freedom)
Lapis lazuli—personal journey
Lodestone—personal journey
Malachite—personal journey, introspection
Moonstone—personal journey
Obsidian—personal journey (insight)

Opal—personal journey (wider vision)
Pearl—personal journey (emotional clarity)
Peridot—personal journey (rebirth/renewal)
Sapphire—personal journey
Serpentine—personal journey
Sodalite—career
Sphene—foundation
Sunstone—personal journey
Tourmaline—personal journey (insight)
Turquoise—personal journey
Zircon—personal journey

Northeast

Agate—wealth
Amazonite—self-cultivation
Amber—wisdom
Amethyst—wisdom
Azurite—wisdom
Calcite—knowledge
Chrysoberyl—luck for attracting wealth
Hematite—self-knowledge, wisdom
Jade—wisdom
Jet—self-cultivation and knowledge
Kunzite—self-cultivation
Labradorite—self-cultivation
Lapis lazuli—wisdom
Onyx—self-cultivation/transformation
Pearl—self-cultivation
Rhodonite—wisdom
Ruby—wisdom
Sapphire—wisdom (mental clarity/intuition)
Sardonyx—wisdom
Sodalite—wisdom
Sphene—knowledge

Sunstone—self-cultivation
Tanzanite—knowledge
Tourmaline—wisdom
Zircon—wisdom

East

Amber—connect with ancestors
Bloodstone—ancestors
Citrine—community (bind relationships)
Coral—community
Diamond—family/community commitment
Herkimer diamond—community (bind relationships)
Jasper—community/family
Rhodonite—community (love of humankind)
Sodalite—community

Southeast

Alexandrite—self-worth and net worth
Bloodstone—attract abundance
Calcite—personal resources
Carnelian—self-worth
Citrine—self-worth, prosperity, wealth
Diamond—wealth, abundance
Emerald—wealth
Hematite—self-worth
Iolite—wealth, resources
Labradorite—resources
Moonstone—wealth
Opal—self-worth
Rhodochrosite—abundance (comfort)
Ruby—wealth, self-worth
Sard—self-worth
Spinel—wealth
Tanzanite—personal resources

Topaz—abundance, self-worth
Zircon—abundance, prosperity

South

Alexandrite—success and reputation
Carnelian—focus for success
Fluorite—illumination
Labradorite—success, reputation
Lodestone—recognition, fame
Malachite—relationships (loyalty)
Opal—illumination (inspiration)
Peridot—success, recognition
Sard—respect, recognition
Sodalite—success

Southwest

Agate—attract love
Amazonite—partner, relationships
Amber—partner/marriage
Aquamarine—relationships, love
Beryl—relationships (communication)
Bloodstone—support in relationships, love
Chrysoberyl—renewal of relationships
Citrine—relationships
Coral—relationships
Diamond—love, relationships
Emerald—relationships
Garnet—partnership
Herkimer diamond—relationships (bind)
Jade—partners, love
Jasper—relationships
Kunzite—relationships
Moonstone—mother, love
Peridot—relationships

Rhodochrosite—love, partner, relationships
Rhodonite—love
Rose quartz—relationships, love
Ruby—love, relationships (loyalty)
Sapphire—love (devotion)
Sard—relationships
Sardonyx—partner (communication), love
Sodalite—relationships
Spinel—relationships
Topaz—love, partner
Tourmaline—partner
Turquoise—partner, love

West

Aventurine—creativity
Calcite—creativity, projects
Carnelian—creativity
Citrine—boosts creativity
Jasper—children
Lapis lazuli—creativity
Moonstone—creativity
Opal—creativity (muse)
Quartz—projects, creativity
Sardonyx—projects
Tourmaline—creativity (inspiration)

Northwest

Calcite—benefactors/helpful people
Garnet—travel
Iolite—helpful people
Moonstone—travel
Peridot—travel
Rhodonite—helpful people (emotional support)
Ruby—father, assisting people

Sapphire—travel
Staurolite—travel
Tanzanite—benefactors

Negative Directions

Agate—protection/combat illness (Misfortune)
Alexandrite—good-luck amulet
Amazonite—disperse negativity
Amber—boost vitality (if kitchen or dining room are in the Misfortune or Loss areas)
Amethyst—protection/good luck
Ametrine—cleansing
Apache tears—protection/luck
Aquamarine—courage/protection
Aventurine—find solutions, make the right choice, luck
Azurite—remove problems through communication, patience
Bloodstone—remove obstacles, neutralize toxins, attract good luck
Calcite—amplify positive energy
Carnelian—soothes energy, protection
Chrysoberyl—protection/luck
Citrine—protection, builds strength, healing
Coral—protection/clearing
Diamond—good-luck amulet
Emerald—banish negative energy and navigate difficulties
Fluorite—good luck, protection during transition
Garnet—personal power to bring "victory"
Hematite—maintain sense of self
Herkimer diamond—cleansing
Iolite—stability, fosters cooperation
Jade—solve problems, bring luck
Jasper—protection against negativity
Jet—protection/calm and cope
Kunzite—remove obstacles

Labradorite—builds vitality (use when negative direction is in kitchen, dining room, or bedroom)
Lapis lazuli—protection
Malachite—protection/banish negativity (Setback and Difficulty)
Moonstone—good fortune
Obsidian—protection/block negative energy
Onyx—protection
Opal—symbol of hope
Peridot—protection/good luck
Rhodochrosite—offsets negativity with love
Rhodonite—banish chaos with calm
Ruby—protection
Sard—protection
Sodalite—resolve issues
Sphene—ability to think things through
Spinel—overcome obstacles (Setback), make amends, protection
Staurolite—protection/luck
Sunstone—deal with fear
Tanzanite—deal with change/weather problems (Difficulty)
Tourmaline—neutralize negative energy, protection
Turquoise—protection

General

Alexandrite—use where healing is needed
Amber—boosts change
Amethyst—manifests change, general healing
Andalusite—manifests change
Aventurine—emotional lift
Benitoite—emotional understanding
Beryl—healing, acceptance
Chrysoberyl—attract luck
Diamond—Longevity direction

Fluorite—boosts vitality
Herkimer diamond—move and raise energy
Iolite—stability
Jade—positive directions Life and Longevity
Kunzite—emotional support
Lodestone—motivation, guidance
Malachite—achieve goals
Onyx—transformation in any life aspect
Peridot—good luck, vitality
Quartz—amplify strengths, transformation
Rhodochrosite—love and caring, support for transitions
Rose quartz—emotional healing
Sapphire—healing
Sard—general healing
Spinel—healing, deal with sorrow/grief, protection
Staurolite—luck
Sunstone—relieve stress
Topaz—move energy, alleviate tension
Turquoise—healing

Glossary

Amulet—A stone or other object engraved with a meaningful symbol that emanates its power to the wearer. Also known as a talisman.

Asterism—The star-like dispersion of light due to tiny "needles" arrayed in several directions within a crystal.

Axis—An imaginary line that passes through a crystal around which its faces are arranged.

Bagua—The octagonal symbol containing the eight trigrams of the *I Ching* that represent the primal or natural energies that shape the world.

B.C.E.—Before Common Era; a nonreligious method for indicating dates before the year 1. B.P., Before Present, is also used.

"Cat's eye"—The technical name is *chatoyancy.* This is caused by the dispersion of light due to tiny parallel "needles" within a crystal.

Chakra—The seven energy centers of the body.

Chi—The life-force energy that flows in all things.

Compass School—*See* Pa Kua Lo Shu School.

Cradle of Mother Earth—This is a configuration—higher terrain behind and an open view in front—that is created when a structure is gently enveloped by and in harmony with the landscape. The palace of Knossos in Crete, the temple at Delphi, Stonehenge, and many ancient sites show this careful placement. This configuration is very similar to the landscape suggested by the Chinese celestial guardian animals utilized by the Form School.

Cryptocrystalline—A crystal structure that is too small to be seen with the naked eye.

Dichroism—Pleochroism in two directions. A stone will appear one of two colors depending on from which direction it is viewed.

Elements—Traditional feng shui utilizes five elements: wood, earth, fire, water, and metal. The elements embody the types of archetypal energy that shapes all things and symbolize the process of change. The interactions among the elements produce a continual cycle of growth and decay.

Energy Group—A group containing four directions. The East energy group consists of East, North, South, and Southeast. The West energy group consists of West, Northwest, Northeast, and Southwest. Each energy group indicates a person's power points or auspicious directions.

Former Heaven Sequence—A configuration of the eight trigrams where they are positioned to represent a perfect universe.

Form School—A feng shui discipline that utilizes landforms and animal symbols in assessing and working with energy.

Hexagram—A combination of two trigrams of the *I Ching*. Set pair of lines represents heaven, man, and earth, and represents a particular pattern or condition.

I Ching—*The Book of Changes;* ancient Chinese book of philosophy and divination. It utilizes sixty-four hexagrams that represent all combinations of human condition.

Kua number—*See* Lo shu number.

Latter Heaven Arrangement—A configuration of the eight trigrams where they are positioned to represent the cycle of the year.

Lo Shu Grid—A three-by-three square grid—the Magic Square—that contains an arrangement of nine numbers that equal fifteen, regardless which direction they are added. Also called the *River Lo Map.*

Lo shu number—Based on a person's year of birth and gender, it is used in conjunction with the *pa tzu* compass to determine one's personal direction and elemental energy.

Lou pan—The basic reference tool of the Compass School of feng shui. It consists of concentric rings that refer to the trigrams, elements, directions, and other attributes. A magnetic compass is located at its center.

Luster—The method by which light is reflected from the surface of a gemstone.

Magic Square—*See* Lo Shu Grid.

Magma—Molten rock that is not stationary.

Pa kua—*See* Bagua.

Pa Kua Lo Shu School—A feng shui discipline that bases its methods on a combination of the eight-sided *pa kua* and the *Lo Shu Grid.*

Pa tzu—A simplified version of the *lou pan* compass containing a trigram, element, *lo shu* number, and energy group identification for each of the nine directions. In Wiccan/Pagan feng shui it is called the *Wheel of the Year compass.*

Pleochroism—A phenomenon caused by the crystalline structure whereby a gemstone will appear as different colors when viewed from different directions.

Piezoelectric—A property of some minerals whereby they release an electric charge when pressure is applied.

Poison arrows—Sharp, straight, destructive energy. Also called *shars.*

Power direction—The direction from which an individual can draw personal strength. It is determined by locating one's *lo shu* number on the *pa tzu* compass. The element related to the power direction holds significant power as well.

Power points—Three directions (in addition to a person's power direction) from which one draws personal power. It is determined by locating one's *lo shu* number on the *pa tzu*/Wheel of the Year compass where an energy group is referenced.

Pyroelectric—A property of some minerals whereby they release an electric charge when heated or cooled.

Refraction—The bending of a light wave within a gemstone.

River Lo Map—*See* Lo Shu Grid.

Scrying mirror—A black, highly polished mirror used in place of a crystal for scrying or crystal gazing. This is performed by focusing one's eyes and attention on the mirror until a trance-like state is achieved whereby one is open to divination.

Secret arrows—*See* Poison arrows.

Sha* or *Shar chi—Negative, unbalanced, and disruptive energy. It is also referred to as the killing breath or poison arrows.

Shars—*See* Poison arrows.

Sheng chi—Positive energy where yin and yang are in balance. It is also called the *dragon's cosmic breath*.

Talisman—A stone or other object engraved with a meaningful symbol that emanates its power to the wearer. Also known as an amulet.

Three As—The three steps to take when performing feng shui: awareness of problems, adjustment to protect against negative energy, and activation of positive energy.

Trichroism—Pleochroism in three directions. A stone will appear one of three colors depending on from which direction it is viewed.

Wealth corner—An area of the living room determined by its location in respect to the main entrance of the home.

Wu Xing—In Chinese, the five elements are called *Wu Xing: Wu* is "five" and *Xing* means "to move." This is the significance of the elements—cycles of movement and change.

Yin and Yang—Binding forces that hold the universe together and are present in all things. One does not exist without the other and each contains a little of the other. In order for a person to exist in wholeness, the yin and yang energies within need to be in balance.

Bibliography

Andrews, Ted. *Crystal Balls & Crystal Bowls: Tools for Ancient Scrying and Modern Seership.* St. Paul, Minn.: Llewellyn Publications, 1998.

Baur, Jaroslav, and Vladimir Boušk. *A Guide in Color to Precious & Semi-precious Stones.* Secaucus, N.J.: Chartwell Books, Inc., 1989.

Berger, Ruth. *The Secret Is in the Rainbow: Aura Interrelationships.* York Beach, Maine: Samuel Weiser, Inc., 1986.

Bowman, Catherine. *Crystal Awareness.* St. Paul, Minn.: Llewellyn Publications, 1997.

Chaucer, Geoffery. *The Canterbury Tales.* Reprint, London: Penguin Books, 1988.

Cunningham, Scott. *Cunningham's Encyclopedia of Crystal, Gem & Metal Magic.* St. Paul, Minn.: Llewellyn Publications, 2001.

Dolfyn. *Crystal Wisdom: Spiritual Properties of Crystals and Gemstones.* Oakland, Calif.: Earthspirit, Inc., 1989.

Govert, Johndennis. *Feng Shui: Art and Harmony of Place.* Phoenix, Ariz.: Daikakuji Publications, 1993.

Graham, Lanier. *Goddesses in Art.* New York: Artabras/Abbeville Publishing Group, 1997.

Heaps, Willard A. *Birthstones.* New York: Hawthorn Books, Inc. 1969.

Hepker, Steven. "Tried and True: Old Standbys Hold Their Own on Drugstore Shelves." *The Star-Ledger* (Newark, N.J.), 30 January 2001, 41.

Jangl, Alda Marian, and James Francis Jangl. *Ancient Legends of Gems and Jewels.* Coeur D'Alene, Idaho: Prisma Press, 1989.

Judith, Anodea. *Wheels of Life: A User's Guide to the Chakra System.* St. Paul, Minn.: Llewellyn Publications, 1996.

Kunz, George Frederick. *The Curious Lore of Precious Stones.* Philadelphia: J. B. Lippincott Company, 1913.

Lin, Jami, comp. and ed. *Contemporary Earth Design: The Feng Shui Anthology.* Miami: Earth Design, Inc., 1997.

Mercer, Ian F. *Crystals.* Cambridge, Mass.: Harvard University Press, 1990.

Pollack, Rachel. *The Body of the Goddess.* Rockport, Mass.: Element Books Limited, 1997.

Robinson, George W., Ph.D. *Minerals: An Illustrated Exploration of the Dynamic World of Minerals and Their Properties.* New York: Simon and Schuster, 1994.

Rossbach, Sarah. *Feng Shui: The Chinese Art of Placement.* New York: Penguin, 1991.

Service, Alastair, and Jean Bradbury. *The Standing Stones of Europe: A Guide to the Great Megalithic Monuments.* London: George Weidenfeld & Nicholson Limited, 1997.

Sofianides, Anna S., and George E. Harlow. *Gems & Crystals from the American Museum of Natural History.* New York: Simon and Schuster, 1990.

Spear, William. *Feng Shui Made Easy: Designing Your Life with the Ancient Art of Placement.* New York: HarperCollins, 1995.

Starhawk. *The Spiral Dance.* San Francisco: HarperSanFrancisco, 1989.

Streep, Peg. *Sanctuaries of the Goddess: The Sacred Landscapes and Objects.* Boston: Bulfinch Press, 1994.

Too, Lillian. *Basic Feng Shui.* Adelaide, Australia: Oriental Publications, 1997.

———. *The Fundamentals of Feng Shui.* Boston: Element Books, Inc., 1999.

Webster, Richard. *Feng Shui for Beginners: Successful Living by Design.* St. Paul, Minn.: Llewellyn Publications, 1999.

White, John Sampson. *Minerals and Gems.* Washington, D.C.: Smithsonian Institution Press, 1991.

Woodward, Christine, and Roger Harding. *Gemstones.* London: The Natural History Museum, 1988.

Wu Xing (members: Joanne O'Brien, Martin Palmer, Eva Wong, Zhao Xiaomin). *The Feng Shui Workbook.* Boston: Charles E. Tuttle Co., Inc., 1998.

Index